Unit Cohesion and Warfare in the Ancient World

This book explores unit cohesion in ancient armies, and how this contributed to the making of war in the Mediterranean world. It takes a varied approach to the subject, from looking at individual groups within larger armies to juxtaposing vertical and horizontal types of cohesion, providing a more detailed understanding of how groups were kept together.

Within the broader definition of 'unit cohesion', this volume approaches more specific aspects of military cohesion in the ancient Mediterranean world including how individual soldiers commit to one another; how armies and units are maintained through hierarchy and the 'chain of command'; and social cohesion, in which social activities and aspects of social power help bind an army or unit together. Examples from across the ancient Mediterranean are explored in this volume, from Classical Greece to Late Antiquity, with topics such as how armies and units cohere during the sacking of cities, Roman standards as a focus of religious cohesion, and how the multi-ethnic mercenary armies of Carthage cohered. Modern approaches to social cohesion are deployed throughout, and these essays serve as an important complement to existing literature on unit cohesion more generally.

Unit Cohesion and Warfare in the Ancient World is of interest to students and scholars of ancient warfare, military history, and military studies, as well as those working on the ancient Mediterranean world more broadly.

Joshua R. Hall is a part-time faculty member at Linn-Benton Community College in Albany, OR, USA. He studies war and social formation in the ancient world, and is the author of *The Armies of Carthage* (Pen & Sword).

Louis Rawlings is a senior lecturer in ancient history at Cardiff University. He has published extensively on Carthaginian, Greek, and Roman history. He is the author of *The Ancient Greeks at War* (University of Manchester Press).

Geoff Lee is an independent researcher, and holds a MA in Ancient History and Classical Studies. He is an organiser of the International Ancient Warfare Conference. He is the editor (with H. Whittaker and G. Wrightson) of *Ancient Warfare: Introducing Current Research, Volume 1* (Cambridge Scholars Publishing).

Unit Cohesion and Warfare in the Ancient World

Military and Social Approaches

Edited by Joshua R. Hall, Louis Rawlings, and Geoff Lee

Routledge
Taylor & Francis Group
LONDON AND NEW YORK

First published 2023
by Routledge
4 Park Square, Milton Park, Abingdon, Oxon OX14 4RN

and by Routledge
605 Third Avenue, New York, NY 10158

Routledge is an imprint of the Taylor & Francis Group, an informa business

British Library Cataloguing-in-Publication Data
A catalogue record for this book is available from the British Library

ISBN: 978-1-138-04585-9 (hbk)
ISBN: 978-1-032-42624-2 (pbk)
ISBN: 978-1-315-17175-3 (ebk)

DOI: 10.4324/9781315171753

Typeset in Times New Roman
by MPS Limited, Dehradun

Contents

Contributors

Adam O. Anders holds a PhD from Cardiff University, specializing in the ancient Roman army. He currently teaches at a private high school in Poland and spends much of his time writing.

Gabriel Baker teaches history at the University of Chicago Laboratory School. He is the author of various articles on ancient warfare, and of *Spare No One: Mass Violence in Roman Warfare* (Rowman & Littlefield).

Ben Greet holds a PhD from the University of Leeds. His research has generally focused on the eagle as a symbol in Roman culture.

Roel Konijnendijk is a lecturer at New College, Oxford. He is the author of many articles on Greek warfare, as well as *Classical Greek Tactics: A Cultural History* (Brill).

C. W. Marshall is a professor of Greek at the University of British Columbia. He is an expert in Greek drama, having written extensively on the topic. His most recent book is *Aeschylus: Libation Bearers* (Bloomsbury).

Aimee Schofield holds a PhD from the University of Manchester. Her research has focused on siege equipment in the ancient world.

Conor Whately is an associate professor at the University of Winnipeg. He has published on many aspects of ancient warfare, and has recently authored *Procopius on Soldiers and Military Institutions in the Sixth-Century Roman Empire* (Brill).

Unit Cohesion in the Ancient World: An Introduction

Joshua R. Hall

What makes soldiers fight? This has been a perennial question for military theorists. Explanations have ranged in perspective from generals who were concerned with the overall function of their armies, groups whose focus was on their own success, to individuals with their personal motivations. This volume uses the concept of 'unit cohesion' to explore this question, and to consider specifically what makes armies, units, and soldiers effective in war and combat. 'Unit cohesion' is a broadly defined set of motivations and pressures that act within armies, units, and between individuals, and which allows for the coordination of soldiers. It affects the effectiveness of military units, whether they are ad hoc agglomerations formed for specific operations, or long-existing entities with formal organizations and institutional frameworks. The former can be as unstructured as a posse or a lynch mob, while the latter may have a long history and strict hierarchy. Loosely organized units can be seen in some of the armies of Carthage, as discussed below by Hall and Rawlings, in which groups of mercenaries were hired for specific campaigns and may not have existed as single, cohesive, units before this. More structured units can be found in armies like that of the Roman Empire, whose institutions, such as the system of standards, discussed in this volume by Ben Greet, contributed to their overall effectiveness.

The types of motivation underlying the concept of unit cohesion are most evident – and important – in combat situations. The scholars who contributed to this volume engage with this important concept in a variety of ways to analyse this fundamental dynamic in the military effectiveness of ancient armies. Chapters range from looking at small units, such as the mercenary companies in service of Carthage, analysed by me and Louis Rawlings, to entire armies, as by Adam Anders in his examination of how battlefield communications contributed to the cohesiveness of the Roman army. By taking various approaches, this book contributes to our understanding of both unit cohesion and ancient armies from many different perspectives and across geographical and chronological boundaries.

It is worth noting that cohesion is not an entirely new concept but has been implicit in 'top-down' explanations that ascribe an army's success to its commanders, discipline, and training. These factors are cited as giving

DOI: 10.4324/9781315171753-1

soldiers the ability to face the dangers and horrors of the battlefield. In the study of Greek warfare, for example, it has often been postulated that the presence of generals on the battlefield helped to motivate their men to fight.[1] For his performance in the Battle of Olpae (426 BC), Joseph Roisman once described Demosthenes' qualities of good generalship as his ability to establish authority over a mixed group of soldiers and effectiveness in utilizing each unit's proficiencies, things that could only be accomplished by a leader on the field.[2] As he described in his overview of Greek generalship, however, the Hellenes believed that an effective general had to be a well-rounded individual, and that battlefield leadership was only one part of a larger personality.[3] Such thinking, to an extent, reflects the emphasis of ancient sources that put commanders at the centre of events (for example, in Xenophon's *Anabasis* or Polybius' *Histories*). There is certainly a kernel of truth to this, and we need only look to the Greek heroic ideal of the warrior and leader to find its origin (or reflection). Similarly, the presence of Roman generals on the field was also portrayed in ancient accounts as a means of enhancing cohesion. Caesar's own narrative of his campaigns is perhaps the most polished example of the 'eye of command' conception. But the supposed self-sacrifice of multiple commanders from the Mus family, whose actions in combat rallied their armies, emphasise the leader's personal contribution in Republican warfare (Livy 8.9, 10.28; Cic. *Fin.* 2.61; *Tusc.* 1.89), and show the pervasive focus on an elite, command, perspective in our surviving accounts.[4] This is reflected in Cicero's belief that true generals were 'those men who are intellectually and theoretically masters' of warfare, and successful in its execution (*De or.* 1.210).[5] His examples are men like Scipio Africanus, Fabius Maximus, Epaminondas, and Hannibal. In this literary setting, it is upon the shoulders of these leaders that victory hangs.

There were also those in the ancient world who saw a connection between civic identity and an army's cohesion. Polybius is the most obvious voicer of this opinion. In his constitutional digression in Book 6, he opined that the armies of Carthage were not as good as those of their Roman opponents because the former consisted of mercenaries rather than people defending their own families and lands (Polyb. 6.52).[6] This was not a unique view, though, and persisted into the modern period. There is still a bias against all-things mercenary, with the word itself being something of a slur in English. This may have been the case in the ancient world.[7]

Others proposed broader conceptualizations of what made armies function. As Xenophon supposedly said in a speech to the Ten Thousand,

> For you understand, I am sure, that it is neither numbers nor strength which win victories in war; but whichever of the two sides it be whose troops, by the blessing of the gods, advance to the attack with stouter hearts, for against those troops their adversaries generally refuse to stand.
>
> (Xen. *An.* 3.1.42)[8]

Spirit, as we would think of it, can be found throughout ancient literature as a motivating factor in warfare. For instance, in his poetic harangue to Spartan soldiers, Tyrtaeus tells them to 'Keep grand and valiant spirits in your hearts / be not in love with life – the fight's with men' (frg. 10 West).

For some modern commentators on the ancient world, comradery and what we would today call 'horizontal cohesion' was one of the most important elements of cohesion (discussed in the next section).[9] Victor Davis Hanson, for instance, in his now controversial description of Greek warfare, believed that 'the comraderie in the Greek ranks' brought a 'unique cohesiveness' to the phalanx formation, and that it was key to the Hellenic victories over the Persians (for instance).[10] Even if this was a significant contributor to the strength of Greek armies, it was not unique to them in the ancient world. The familial nature of early Italic warbands is an interesting example of a military structure that relied extensively on this type of cohesion.[11] It also played a part in the functioning of the Sacred Band of Thebes and, beyond antiquity, in institutions such as the 'warrior fraternities' of the Pueblo Nations in the American Southwest.[12]

There were certainly different types of cohesion at work in ancient armies, some of which were picked up on by authors from the period, as we have seen in brief above. But the concept of 'unit cohesion' has been much refined in modern military studies, as will be seen below, and examining ancient armies through this lens is our purpose in the following chapters.

Unit Cohesion

Modern scholars have refined the discussion of combat motivations by considering 'unit cohesion' as an umbrella term for a complicated and debated set of forces, which are exerted within military groups, small and large. It is usually further broken down into 'vertical' and 'horizontal' cohesion. The former 'refers to downward or upward cohesion involving leaders and followers', connected in the hierarchy of command, while the latter often 'refers to cohesion at the primary group level – generally the crew or squad, and perhaps the platoon'.[13] Is it worth noting here that 'horizontal' also can apply to the analysis of larger units such as regiments, especially where smaller groups are not organizationally articulated. As will be discussed below, such a definition would include many ancient formations such as hoplite *lochoi* or tribal *taxeis*, or legionary maniples and cohorts, which numbered in the hundreds rather than the tens. Horizontal cohesion can also be 'secondary', that is the inter-relationship between similar sized units within a larger organization, such as those that make up a battalion, division, or even wing of an army.

Vertical cohesion should perhaps be seen as the older of the two concepts. The emphasis in the past, including antiquity (as seen above), on the importance of good leadership, strong discipline, *esprit de corps*, and 'patriotism' is evidence of this. In many ways, these all fall under the category of vertical cohesion.[14] As we have seen above, leadership and civic identity were understood to be important by Greek and Roman thinkers. Likewise, within modern research, both have been shown to contribute to overall military cohesion.[15]

Although having roots in earlier thinking, the modern reflection on horizontal cohesion did not fully emerge until the wake of the Second World War. S. L. A. Marshall, a historian in the service of the U.S. Army, wrote in the aftermath of that conflict, that:

> I hold it to be one of the simplest truths of war that the thing which enables an infantry soldier to keep going with his weapons is the near presence or the presumed presence of a comrade. The warmth which derives from human companionship is as essential to his employment of the arms with which he fights as is the finger with which he pulls the trigger or the eye with which he aligns his sights.

He went on to conclude that a soldier is

> sustained by his fellows primarily and by his weapons secondarily. Having to make a choice in the face of the enemy, he would rather be unarmed and with comrades around him than altogether alone, though possessing the most perfect of quick-firing weapons.[16]

In the United States, research into the effects of unit cohesion led to the creation of the COHORT system in 1981. This established the concept of combat units 'based around a cohesive nucleus of soldiers'.[17] COHORT was an acronym: COHesion, Operational Readiness, Training. Cohesion thus became part of the overall training regime for army units, in this case largely of the 'horizontal' type. In a newspaper article, written not long after its implementation in a number of groups, it was noted that the program overall had generated a number of positives: 'retention is up, discipline problems are down, there are fewer racial and ethnic cliques, and greater unit pride is evident'.[18] Two particular activities were noted: when a soldier in one of the trial units was married, their entire platoon attended in addition to other members of their company, and that at Thanksgiving the entire company dined together in uniform. These are examples of practices that could help build horizontal cohesion in the form of peer-bonding. The process of sustained socialization of individuals into a coherent group with a collective sense of identity was observed to work best among relatively small numbers of soldiers who were intimates; it became more diluted in larger organizations and units, where soldiers had less intimate or frequent contact with the other members.

As exemplified by Marshall, intimate social cohesion used to be considered paramount. Even before 'unit cohesion' as a subject of study became mainstream, the 'buddy principle' (as it is sometimes called) was thought to be important. As long ago as 21 October 1798, George Washington concluded a letter to Henry Knox by saying that 'his first wish would be that my Military family – and the whole Army – should consider themselves as a band of brothers, willing & ready, to die for each other'.[19] This concept also can be seen in early twentieth-century training manuals from the United States. In his *Manual of Military Training*, J. Moss wrote that

> comradeship is a very valuable military characteristic. What a world of meaning there is in the words, "Me and my Bunkie." A soldier may have many acquaintances and a number of friends, but he has but one "Bunkie." In times of great danger two men who are "bunkies" will not skirt so easily as two independent men.[20]

Here we find a similar notion to that quoted from Marshall, above, and found in George Washington's letter. This idea was supported by a study from the early years of this century.[21] But the peer-bonds between individuals that were so touted by earlier theorists have been attacked in contemporary work.

A recent meta-study argued that there is no evidence for a causal relationship between social cohesion and effectiveness in combat.[22] Instead, the primary force of horizontal cohesion behind military success is 'task cohesion'.[23] According to this argument, the primary driving force between unit cohesion and military success is the desire to complete specific tasks. This echoes Wm. Darryl Henderson's definition of cohesion in general:

> [it] exists in a unit when the primary day-to-day goals of the individual soldier, of the small group with which he identifies, and of unit leaders are congruent – with each giving his primary loyalty to the group so that it trains and fights as a unit with all members willing to risk death to achieve a common objective.[24]

Similarly, an article written as part of a major psychiatry textbook concludes that 'the more interdependence among the members [of a unit] that is necessary for success, the greater the payoff in cohesion'.[25] In other words, tasks which require the entire unit's cooperation enhance the cohesive forces present therein.

As we have just seen, the concept of 'unit cohesion' is multi-faceted and contested. Its various strands, regardless of whether all modern theorists support them, provide important insights into the *possible* mechanisms by which armies function(ed). The purpose of this book is to look at aspects of ancient warfare through the different lenses of unit cohesion. Cohesion remains a hot topic and, in recent years, has moved from its traditional Eurocentric and modern perspective, to recognize the significance of non-Western and pre-twentieth century exempla.[26] This volume is further intended to contribute to this widening debate by looking at a range of ancient militaries that rarely considered by modern theorists.

Cohesion in Ancient Armies

Until now, few scholars have attempted to apply modern research on unit cohesion to the armies of the ancient Mediterranean world.[27] The essays in this book aim to fill this gap investigating specific aspects of war from Classical Greece to Late Antiquity. Our objective was not to be comprehensive, which is probably impossible in a single volume of this length, but rather to offer a variety of studies on different situations where cohesion contributed to the functioning of ancient armies. In this way, historians of various periods with an interest in warfare will find something of value. Specialists will also find individual chapters related to their research helpful. We also hope that these essays will encourage further study into the forces of cohesion in pre-modern military units.

The chapters of this book are organized in a roughly chronological order. It opens with two discussions of Classical Greece. Roel Konijnendijk tackles the question of how Athenian hoplite armies cohered. He explores the problems inherent to non-professional armies, noting issues such as the impact of having no training when trying to execute battlefield manoeuvres. Although there were strong horizontal forces of cohesion at work in Athenian levies, notably commitment to the primary group and the values and goals shared by members of these armies, the lack of structured vertical forces was a problem. Konijnendijk deftly ranges through these aspects of the ancient Greek phalanx, helping to further our understanding of how and why hoplites fought. In the next chapter, C. W. Marshall focuses on a specific force, the Ten Thousand, as described in Xenophon's *Anabasis*. In contrast to Konijnendijk, this chapter singles out a group based on their armament: slingers. Not only is the importance of ad hoc units of slingers emphasized, but also their role in increasing the morale of the army and in functioning to strengthen cohesive forces within it.

Aimee Schofield provides the first of two chapters that look at siege warfare. She specifically examines the text of Aeneas Tacticus and how unit cohesion is represented in it. The concept of *homonoia* – unanimity – is the foundation of both Aeneas' and wider Greek thoughts on siege warfare and how *poleis* were to be defended. Schofield identifies the key differences between fighting in the field and defending a city, which affected cohesion, including the presence of families and the problematic socio-political relationships present in any state. Gabriel Baker follows this by looking at cohesion in the forces that breach a city's fortifications. By taking a broad definition of cohesion as 'collective combat performance' and the ability 'to act together and to achieve their mission in the face of the enemy'[28] – he argues that specific tasks undertaken after the breach and inside the walls helped to focus soldiers' concentration, and to maintain cohesion. Importantly, Baker also notes that, while sometimes soldiers could run amok, vertical cohesion was as often maintained and that commanders had various ways to keep control of their armies.

Chapter five again shifts our focus, this time to the armies of Carthage. Louis Rawlings and I look at how these functioned and cohered, both vertically and horizontally. Through a detailed examination of the evidence, we argue for a new view of Punic armies that discards old biases about the polyglot and non-national aspects. What emerges is an image of strong, well-functioning military units not dissimilar from those of other peoples of the Mediterranean, including Rome.

Adam Anders and Ben Greet each discuss aspects of unit cohesion in Roman armies. The first focuses on communication as a tool for both battlefield management and as a means of maintaining cohesion. Anders concludes that various means of transmitting commands were essential for maintaining vertical cohesion, and operated not only between generals and their troops, but also between subordinate officers and their men; the role of tribunes is especially highlighted. Greet takes a different approach, examining the social cohesion imparted by legionary standards. By studying the significance of these, their meaning, their position in the army and camps, and cult around them, he concludes that the standards were infused into personal beliefs, thus fostering even stronger peer bonds amongst the men of the legions.

Conor Whately's chapter on the armies of late antiquity and Louis Rawlings' on the breakdown of cohesion close out this volume. The former's focus is on horizontal cohesion in small units of the Roman army between AD 491 and AD 602. Surveying a wide range of evidence (including epigraphy, material culture, and literature), he concludes that there is too little information to suggest that primary-group peer bonding was a decisive factor in the armies of this period. His results echo some of the modern literature noted above, which argues that peer bonds are not nearly as important as has sometimes been believed. Rawlings' chapter examines the breakdown of cohesion and its causes. He begins by looking at predispositional factors – those that exist in a unit before combat begins – such as insubordination and disaffection. He also highlights what some readers may find to be paradoxical, that deploying units of both inexperienced and veteran soldiers can undermine cohesion. The analysis continues by discussing battlefield factors that contribute to the breakdown of cohesion, such as casualties and the mental stresses of combat itself. Rawlings closes his chapter by pointing out the persistence of cohesion, even during rout. Overall, this provides a helpful balance to the other chapters in the volume, and lets readers see how the concepts discussed in them can falter.

As readers will be able to tell from this summary, these chapters cover a wide temporal range and approach unit cohesion from various angles. One of our goals was to not restrain authors to the editors' conception of this phenomenon, nor to limit them to a single chronological period. What this has achieved is a diverse set of ideas and room for further debate on the importance of cohesion in ancient armies.

Notes

1 e.g. Hanson (1989): 107–16; Wheeler (1991); Pritchett (1994a); idem. (1994b).
2 Roisman (2017): 5.
3 Roisman (2017): 7–14.
4 Though the Roman explanation was that it garnered the army religious favour, of course, Versnel (1981). On the importance of the 'eye of command' perspective, see Kagan (2006).
5 See Hall (2019). Granted, Elizabeth Rawson warned us long ago that 'no one of course should look for lofty critical standards to be exercised on the [exempla]' of Cicero, Rawson (1972). So perhaps we should exercise caution when trying to extract general conceptions.
6 Louis Rawlings and I argue below that this was not the case. Other examples of internal cohesion can be found, such as in the army of Wellington during the Peninsular War, see Coss (2005).
7 Trundle (2004): 19–21.
8 Perhaps, though, better read as an early version of Voltaire's 'Dieu n'est pas pour les gros bataillons, mais pour ceux qui tirent le mieux'.
9 See also MacCoun and Hix (2010): 139–41.
10 Hanson (1989): 117. This should be tempered with more recent readings of the evidence such as Crowley (2012) and Konijnendijk (2017).
11 Armstrong (2016).
12 Walker (2009): 123–4.
13 McCoun and Hix (2010): 139–40.
14 See, for instance, the definition and description of cohesion in Castillo (2014): 17–19.
15 On nationalism, see now Castillo (2014).
16 Marshall (1947, repr. 2000): 42–3.
17 van Epps (2008): 103.
18 Knickerbocker (1982).
19 *Founders Online*: http://founders.archives.gov/documents/Washington/06-03-02-0087. The notion that a military unit should be viewed as a family analogue is found throughout the literature, note for instance Henderson (1985): 14: 'Primary social affiliation within the unit is an extremely significant indicator of cohesion because it means that the small military unit has replaced other influences such as the family as the primary determinant of the soldier's day-to-day behavior'. In the ancient world, a unit could actually consist of family members; Armstrong (2016).
20 Moss (1917): 252.
21 Wong *et al.* (2003).
22 MacCoun *et al.* (2005). Note the critique of Siebold (2007) and (2011).
23 See also MacCoun and Hix (2010): 142–3. Note, however, the objections of Siebold (2015). This volume has followed current usage to understand task cohesion as Henderson (1985) defines it and as Armstrong (2016) applies it to Roman warfare.
24 Henderson (1985): 4; idem (1979): 3–18.
25 Manning (1994): 9.
26 Käihkö (2018). Note the critique by King (2021) and Käihkö's response (2021).
27 Notable exceptions are Crowley (2012) and Armstrong (2016), the latter of whom used the framework to discuss warbands in early Rome.
28 Citation in Baker's chapter in this volume.

Bibliography

Armstrong, J. 2016: 'The Ties that Bind: Military Cohesion in Archaic Rome'. In J. Armstrong (ed.), *Circum Mare: Themes in Ancient Warfare*, Leiden: 101–119.

Castillo, J. J. 2014: *Endurance and War: The National Sources of Military Cohesion*, Stanford.

Coss, E. J. 2005: *An analysis of the campaign and combat experiences of the British soldier in the Peninsular War, 1808–1814*. Unpublished PhD Thesis, The Ohio State University.

Crowley, J. 2012: *The Psychology of the Athenian Hoplite: The Culture of Combat in Classical Athens*, Cambridge.

Hall, J. R. 2019: 'A hero from an earlier age: Epaminondas in Cicero and Roman antiquity', *Ancient World Magazine*, 21 May, 2019: https://www.ancientworldmagazine.com/articles/hero-earlier-age-epaminondas-cicero-roman-antiquity/

Hanson, V. D. 1989: *The Western Way of War: Infantry Battle in Classical Greece*, second edition, Berkeley.

Henderson, Wm. D. 1979: *Why the Vietcong Fought: A Study of Motivation and Control in a Modern Army in Combat*, Westport, CT.

Henderson, Wm. D. 1985: *Cohesion: The Human Element in Combat. Leadership and Societal Influence in the Armies of the Soviet Union, the United States, North Vietnam, and Israel*, Washington, DC.

Kagan, K. 2006: *The Eye of Command*, Ann Arbor.

Käihkö, I. 2018: 'Broadening the Perspective on Military Cohesion', *Armed Forces & Society* 44 (4): 571–586.

Käihkö, I. 2021: 'Toward Strategic Cohesion: A Reply to King's Criticism of the Call for a Broader View of Cohesion', *Armed Forces & Society* 47 (3): 596–603.

King, A. 2021: 'Broadening the Perspective on Military Cohesion? A Reply', *Armed Forces & Society* 47 (3): 586–595.

Knickerbocker, B. 1982: 'Army's COHORT plan keeps units together, builds morale', *The Christian Science Monitor*, December 22, 1982.

Konijnendijk, R. 2017: *Classical Greek Tactics: A Cultural History*, Leiden.

MacCoun, R. J. *et al.* 2005: 'Does Social Cohesion Determine Motivation in Combat? An Old Question with an Old Answer', *Armed Forces & Society* 32 (4): 646–654.

MacCoun, R. J. and W. M. Hix, 2010: 'Unit Cohesion in Military Performance'. In B. D. Rostker *et al.* (eds.), *Sexual Orientation and U.S. Military Personnel Policy: An Update of RAND's 1993 Study*, Santa Monica: 137–165.

Manning, F. J. 1994: 'Morale and Cohesion in Military Psychiatry'. In F. D. Jones *et al.* (eds.), *Textbook of Military Medicine, Part I: Military Psychiatry: Preparing in Peace for War*, Falls Church, VA: 1–18.

Marshall, S. L. A. 1947 (repr. 2000): *Men against Fire: The Problem of Battle*, Norman, OK.

Moss, J. A. 1917: *Manual of Military Training*, second, revised edition, Menasha, WI.

Pritchett, W. K. 1994a: 'The General on the Battlefield'. In W. K. Pritchett (ed.), *Essays in Greek History*, Amsterdam: 111–144.

Pritchett, W. K. 1994b: 'The General's Exhortation in Greek Warfare'. In W. K. Pritchett (ed.), *Essays in Greek History*, Amsterdam: 27–109.

Rawson, E. 1972: 'Cicero the Historian and Cicero the Antiquarian', *Journal of Roman Studies* 62: 33–45.

Roisman, J. 2017: *The Classical Art of Command. Eight Greek Generals Who Shaped the History of Warfare*, Oxford.

Siebold, G. L. 2007: 'The Essence of Military Group Cohesion', *Armed Forces & Society* 33 (2): 286–295.

Siebold, G. L. 2011: 'Key Questions and Challenges to the Standard Model of Military Group Cohesion', *Armed Forces & Society* 37 (3): 448–468.

Siebold, G. L. 2015: 'The Misconceived Construct of Task Cohesion', *Armed Forces & Society* 41 (1): 163–167.

Trundle, M. 2004: *Greek Mercenaries: From the Late Archaic Period to Alexander*, London.

Van Epps, G. 2008: 'Relooking Unit Cohesion: A Sensemaking Approach', *Military Review* 88 (6): 102–110.

Versnel, H. S. 1981: 'Self-sacrifice, Compensation and the Anonymous Gods'. In Rudhardt, J. and Reverdin, O. (eds.), *Le sacrifice dans l'antiquité*, Geneva: 136–185.

Walker, W. H. 2009: 'Warfare and the Practice of Supernatural Agents'. In A. E. Nielsen and W. H. Walker (eds.), *Warfare in Cultural Context: Practice, Agency, and the Archaeology of Violence*, Tuscon: 109–135.

Wheeler, E. L. 1991: 'The Hoplite as General'. In V. D. Hanson (ed.), *Hoplites: The Ancient Greek Battle Experience*, London: 121–171.

Wong, L. *et al.* 2003: *Why They Fight: Combat Motivation in the Iraq War*, Carlisle Barracks, PA.

1 The Eager Amateur: Unit Cohesion and the Athenian Hoplite Phalanx

Roel Konijnendijk

How did a Classical Greek hoplite brave the horrors of hand-to-hand combat? Given the central role played by John Keegan's 'Face of Battle' approach in the late twentieth-century revitalisation of the academic study of Greek warfare, it is somewhat surprising that until recently this matter was not examined in detail. Scholars seem to have found it sufficient to point in the general direction of comradeship, social control and shared experience to explain Greek warriors' will to fight. Rarely was there any in-depth analysis of the origin or nature of hoplite unit cohesion and fighting spirit.[1] With *The Psychology of the Athenian Hoplite,*[2] Jason Crowley addressed this peculiar shortcoming in exemplary fashion. He was the first to apply modern theories of combat effectiveness to Classical Greece in a systematic way; his work laid bare the military structures, martial values, and battle sociology – in short, the 'culture of combat' – that produced the Athenian hoplite.

Crowley's aim was to explain this warrior's solid record of victories in battle despite a low degree of organisation and typically unsophisticated tactical behaviour.[3] His conclusions are very positive. The absence of training camps to inculcate warriors with military values and skills was made up for by the ubiquitous presence of warlike themes, attitudes, and symbols in the rituals, traditions, art, and architecture of the polis. The militia's ad hoc nature was balanced out by the fact that hoplites fought side by side with their neighbours and next of kin, bringing their existing primary groups to the battlefield with them. Commanders' lack of professional military education was outweighed by their experience and by the trust the hoplites revealed they had in them when they elected them in the first place. Most importantly, Athenian hoplites' moral commitment to their city's cause and to warfare as a test of their worth as citizens allowed them to face the ordeal of close combat whenever they were called upon to do so.

This reading is compelling, but also remarkably optimistic. For all his constructive efforts, Crowley presents a considerable list of things the Athenian militia had to do without. Hoplites lacked even the most basic training, and consequently had no tactical abilities worthy of the name. Their commanders were amateurs, their units large and unwieldy, and the officer level that modern theorists regard as key to military compliance relationships

DOI: 10.4324/9781315171753-2

was altogether absent. The hoplites' meagre compensation for service was insufficient to act as a monetary incentive, while the lack of a 'coercive apparatus' to secure discipline meant that they could not be forced to obey if they did not want to. Perhaps most damning of all is the absence of permanently established military units comparable to the regiments of the modern British Army, and therefore the absence of the shared identity and traditions that characterise such units. This absence meant that the Athenian hoplite had to fight, in Crowley's words, 'without the benefit of military unit cohesion'.[4]

In my view, these points add up to a somewhat less rosy picture. It is not my intention here to launch an ill-advised assault on Crowley's groundbreaking work; rather, I mean to look at his subject from a different perspective, to see if what we know about the meagre martial capabilities of the Athenian hoplite militia may help us understand the military methods and tactical theory of Classical Athens. Crowley, in effect, argued that the horizontal cohesion of strongly motivated citizen soldiers more than made up for the limited vertical cohesion of the Athenian phalanx.[5] I will suggest instead that even the Athenians themselves increasingly recognised their militia's strong emotional ties and shared sense of purpose as a patch on the hoplites' lack of proper military organisation, training, and discipline. The absence of all the trappings of a professional military meant that their unit cohesion and combat effectiveness hinged in the final instance on individual warriors' moral conviction alone. As a result, the behaviour of Athenian hoplites in combat was unpredictable. By the fourth century, several Athenian thinkers – notably the veteran commander Xenophon – were urging their countrymen to do something about it.

An Amateur Army

To put the question in its appropriate context, it will be worthwhile first to explore the ways in which the Athenian militia fell short of what we might assume an army to be. It is important to acknowledge just how different their approach to military matters was, and how it generated its own unique challenges and priorities.[6]

First of all, despite the reservations of some scholars,[7] Crowley is no doubt correct to assert that 'the Athenian hoplite did not receive any military training, basic or otherwise'.[8] This is confirmed by Xenophon, who states plainly that 'the city does not publicly train for war' (*Mem.* 3.12.5). It is also confirmed by Plato, who claims more broadly that a training programme like the one he envisions does not exist 'in any city-state at all, except maybe in a very small way' (*Leg.* 831b). It is confirmed again by Isocrates, who complains that for all the Athenians' zeal for war 'we do not train ourselves for it' (8.44). The complete lack of evidence for Athenian hoplite formation drill or organised weapons training until the very end of the Classical period shows that these are not mere rhetorical statements.

Crowley argued persuasively that, in a society as steeped in war and martial values as Classical Athens, there was no need for the moral transformation of civilians into soldiers that forms a major part of modern basic training.[9] In that sense, Athenian hoplites' lack of training need not have made them any less effective in combat. They were taught from childhood to idealise warriors and did not object to the use of lethal force against the enemy. They knew that they fought for the survival of their community in a more direct and literal fashion than almost any soldier in a modern professional army. But the training discussed by the Classical authors cited above was not about attitude; it was about physical fitness, military discipline, formation drill, and weapon proficiency.[10] In these areas, the Athenian hoplite received no formal instruction at all.

Modern scholars have often underestimated how much this lack of training would have impeded their tactical abilities. Even Crowley assumes that they would have been capable of some basic manoeuvres, like wheeling or forming into a hollow square.[11] Yet, having never learned any unit drill, it is extremely unlikely that Athenians would have been able to carry out the formation evolutions required to complete such manoeuvres in an orderly fashion. Xenophon asserts that most people, including the self-proclaimed drillmasters of his day, dismissed even the simplest manoeuvres as hopelessly complex (*Lac.* 11.5, 8). Elsewhere, he sketches a humorous picture of an untrained unit's first attempt to march in formation, which goes wrong because the recruits fail to grasp its most elementary features (*Cyr.* 2.2.6–9). Without at least some preparation as a unit, any attempt to manoeuvre in formation would end in chaos.[12]

Xenophon was a keen advocate of the Spartan solution to this problem: as long as the entire first rank of the phalanx consisted of officers and file leaders who knew what they were doing, the men in the other ranks could simply be told to 'follow the man in front' (*Lac.* 11.4–6; *Cyr.* 2.2.8), which surely required little training. In this way, a phalanx of amateurs could be made capable of manoeuvre with minimal effort. However, this method was predicated on the existence of an elaborate officer hierarchy, and a battle formation in which 'the care of what needs to be done falls upon many' (Thuc. 5.66.4). The Athenian phalanx was no such formation. It knew only three officer levels (*strategos*, *taxiarchos,* and *lochagos*), the lowest of which stood at the head of a unit (the *lochos*) of at least one hundred and possibly several hundred men. We hear of no officers subordinate to the *lochagos* in any Athenian army.[13] As a result, even when mustered at its full strength of many thousands, the Athenian hoplite body never contained more than a few dozen men who had 'the care of what needs to be done'. Furthermore, these officers were not professional soldiers themselves. Some may have known very well what needed to be done, but others, for lack of experience or tactical awareness under extreme pressure, may not.[14] The leadership of a few veteran commanders could never have sufficed to show the entire phalanx what was expected of it.

When orders had to be issued, Athenian generals could not rely on this underdeveloped chain of command to pass them on. The only solution was to make commands go over the heads of subordinate officers and reach the men in the ranks directly. The all-important signals to advance and to retreat were not given by voice or by example, but by the unmistakable blast of the *salpinx*.[15] However, since hoplites were not trained like Roman legionaries to recognise a range of trumpet signals, the advance and the retreat were in fact the *only* orders an Athenian commander could give in the expectation that they would be followed.[16]

It is not surprising, then, that the sources do not record a single instance of an Athenian phalanx successfully carrying out a formation evolution. Athenian rank-and-file formations, including the hollow square formed by the reserve at Syracuse (Thuc. 6.67.1), were orderly because they were static. Only their prolonged service together eventually seems to have enabled the veterans of the Sicilian expedition to retain such a formation while in motion (Thuc. 7.78.2). The only exception to the rule that Athenians did not manoeuvre is Cleon's attempt to wheel his right wing away from Amphipolis (Thuc. 5.10.4), which prompted chaos and defeat. He clearly asked too much of his men, and his enemy, the Spartan Brasidas, successfully exploited their confusion. Xenophon's repeated laments about the uselessness of an army marching in disorder ('those who walk hindering those who run, and those who run hindering those who are standing still', *Oec.* 8.4; 'men getting in each other's way like a crowd leaving the theatre', *Eq. mag.* 2.7) suggest that the sight of a hopelessly tangled Athenian army remained all too familiar even in the mid-fourth century.

The absence of other features of modern military organisation also implies limited abilities in combat. The proud standards and traditions of permanently established military units might have done much to enhance the performance of their members in battle and mitigate the lack of organisation and training of the hoplite body. However, as Crowley concluded, such units did not exist in Classical Athens.[17] The Athenian archers and cavalry may have had a claim to being more or less standing units, and the latter was indeed encouraged to develop some degree of unit pride (Xen. *Eq. mag.* 1.22–25, 9.3–4; *Eq.* 10.11–12; *Vect.* 2.5), but the tribal *taxeis* that made up the hoplite phalanx were formed and disbanded entirely at need. Neither the composition nor the commanders of these formations lasted beyond a single campaign, and no cadre remained to preserve an abstract notion of what the unit was or what it stood for. In the absence of such 'military unit cohesion', hoplites could bring to bear only the skills and values they brought with them when they were called up to fight.

The role of junior officers in securing the loyalty and lasting effectiveness of modern military units is widely recognised; the model borrowed by Crowley in his discussion of the compliance relationship – the system that keeps men fighting and obeying orders – assumes that commanders of sections and platoons are central to that relationship. These officers form a bridge between soldier and commander, combining dedication to the mission with personal ties to their men. However, Crowley admitted that he was forced to deviate from the model on this point.[18] As we have seen, there were no small units or small unit commanders in the Athenian army. Crowley's solution was to transplant the crucial roles of the junior officer to higher-ranking hoplite commanders, but this seems overly optimistic. Simply put, there were not enough of them to ensure compliance through association with the men in the ranks. Each commanded hundreds, if not thousands, of men; how many of those could they have personally known and interacted with?[19] Xenophon may have been a proponent of strong ties between generals and their men (*Ages.* 8.1–2; *Hell.* 5.1.3–4; *Mem.* 3.1.9, 3.2.3–4; *Oec.* 21.7–8), but he praises his fictional Cyrus at length for taking 'special effort' to learn the names of just his most prominent officers (*Cyr.* 5.3.46–50). Clearly, we should not imagine that Greek officers of any rank, including even the most personable and inspiring ones, associated closely with a significant number of their subordinates. It seems unlikely that these officers contributed substantially to the hoplites' willingness to fight, except through persuasive speech and personal example.

Finally, as many authors have noted, the Athenian state lacked an effective system to enforce military discipline.[20] Crowley neatly summed up the problem: 'Athens, primarily for ideological reasons, never developed a coercive apparatus capable of forcing unwilling combatants to comply with her demands'.[21] The ideological reasons in question were an aversion to any sort of imposed hierarchy, a deep-seated suspicion of one-man rule – however temporary and circumscribed – and a commitment to consensus-based decision making.[22] We hear of few commanders willing to go against this ideology for the sake of military efficiency. They rightly feared that any misstep in the eyes of their subordinates would see them put on trial on their return to Athens (Thuc. 7.48.3–4).[23] Therefore, while generals theoretically had the authority to fine, discharge, or imprison insubordinate troops ([Arist.] *Ath. Pol.* 61.2), and could even have men summarily executed, they were understandably hesitant to use these powers in practice. The result was that Athenian hoplites could expect to get away with even blatant forms of disobedience or dereliction of duty. There was only the letter of the law to stop them from ignoring orders, fleeing from battle, or even deserting from the army altogether. As Xenophon put it disapprovingly, everywhere except in Sparta, 'when a man proves a coward, the only consequence is that he is called a coward' (*Lac.* 9.4).[24]

In short, the Athenian hoplite militia lacked nearly all of the features that we associate with a military force worthy of the name. They had no combat training, no formation drill, no elaborated chain of command or small unit organisation, no unit traditions, no junior officer cadre, and no enforced disciplinary measures. They received no formal instruction in what they were expected to do, and their leaders were only nominally able to persuade or force them to do it. In these circumstances, the horizontal unit cohesion of the hoplites – their commitment to their primary group, the values it stood for, and the goals it shared – was absolutely essential. It was practically the only thing holding the Athenian phalanx together.

The Alternative

The Greeks themselves were clearly aware of the importance of comradeship and the moral support of peers in maintaining an army's will to fight. Xenophon states outright that 'there is no stronger phalanx than one composed of friends fighting together' (*Cyr.* 7.1.30). In literary sources from Homer (*Il.* 2.362–368) to Onasander (*Strat.* 24) we find the advice to deploy men with existing social relationships side by side in the battle line.[25] Trust in the cohesive power of personal bonds also inspired fourth-century Athenian philosophers' discussion of the potential merits of a military unit composed entirely of pairs of homosexual lovers (Pl. *Symp.* 179e–180b; Xen. *Symp.* 8.32–35).[26] Greek thinking along these lines suggests a general understanding that social cohesion as a form of horizontal unit cohesion was essential to the performance of a militia army.[27]

Indeed, direct verbal encouragement of fellow combatants – perhaps the most 'visible' form that social cohesion takes in the literary record – is a prominent feature of the *Iliad*[28] and is singled out by Tyrtaeus as the mark of 'a man good in war' (fr. 12, lines 19–20). Thucydides claims that the Spartans preferred it over a speech by their commander as a way to encourage the men in the ranks (5.69.2); Xenophon also notes the Spartan practice of having the men exhort each other (*Lac.* 12.9). Following their model, Xenophon's hero Cyrus orders the rear ranks of his army to urge on the men in front, and vice versa (*Cyr.* 3.3.41–42), resulting in a striking visualisation of an advancing phalanx exhorting itself:

> When the paean was over, the Equals marched forward, beaming, moving expertly, checking on one another, calling those beside and behind them by name, repeating, "come on, friends," "come on, brave men," encouraging each other to advance. And those behind, hearing them, in turn called on the men in front to lead on valiantly. In this way Cyrus' army was filled with enthusiasm, ambition, strength, courage, cheering, self-control, obedience ....
>
> (Xen. *Cyr.* 3.3.59)

Yet the context of these last few passages reveals a growing sense that there was more to unit cohesion than the encouraging presence of the primary group. The men described here were not effective only because they were friends; indeed, their friendship was not the key to their military effectiveness.

In these passages, Thucydides and Xenophon describe the mutual encouragement of fellow combatants specifically among Spartans, or forces inspired by the Spartan model.[29] This is an important distinction, since, on the battlefield, Spartans were not like other Greeks. The Spartans alone were thought to have certain abilities that set them above a simple reliance on primary-group support. In Xenophon's treatment of the notion of couples fighting together, he has Socrates argue that the Spartans are brave enough to fight well even when drawn up 'among strangers and not with their lovers' (*Symp.* 8.35). Elsewhere, Xenophon stresses that only Spartiates know how to form up and fight together with anyone who happens to be present (*Lac.* 11.7). The point of both lines was to show that these men had something other Greeks did not – something that allowed them to keep their resolve even when their close companions were absent.

Nevertheless, these authors characterise *Spartan* armies, not ordinary Greek militias, as a uniquely mutually supportive environment. Scenes of men cheering each other on as they advance into battle are not seen in descriptions of non-Spartan forces. It seems that our sources are deliberately presenting the exhortations of peers as no more than a *secondary* reinforcement of unit cohesion, which was itself the result of other, more fundamental factors. In Thucydides' account of the battle of Mantinea – where, as noted above, the Spartans are said to rely on informal exhortation rather than a general's speech – the historian underlines the Spartan army's superior organisation, officer hierarchy, and careful preparation for war (5.66.3–4, 69.2, 70). It was the Spartans' training and organisation that made them believe they could do without a rousing speech. It was that same training and organisation, and not their shouts of encouragement, that allowed them to march into battle calmly and collectedly, and to retain their cohesion after the initial encounter of the lines. Similarly, in his description of Cyrus' advance into battle, just before the passage about the mutual exhortation of the troops, Xenophon establishes the real reasons for the efficiency and high morale of these men:

> [Cyrus] led on quickly, and they followed in good order, for they understood marching in formation and had practised it; moreover, they followed courageously, because they rivalled each other in desire for victory, and because their bodies were well-trained, and because the front-rank men were all officers.
>
> (Xen. *Cyr.* 3.3.57)

The litany of military skills presented by these authors forms the context of their comments on the value of social cohesion. In their accounts, the presence and the encouragements of peers had a positive effect only because they confirmed what the men already knew: that they were well-prepared to meet the enemy, that their formation was in good order, and that their officers were skilful and brave. Their confidence in their own superiority inspired them to urge each other on.

By contrast, when Classical authors comment on the Athenian hoplite body, they often highlight its shortcomings. I have already cited the various passages lamenting its lack of training; Xenophon also lambasts it for its disobedience (*Mem.* 3.5.19) and cowardice (*Symp.* 2.13) while the Old Oligarch claims it had a reputation for weakness ([Xen.] *Ath. Pol.* 2.1). Thucydides, though he sometimes portrays Athenian forces as experienced and reliable (6.70.1, 72.3), also notes how easily they fell into confusion, leading to self-inflicted casualties and defeat at Delium (4.96.3), and further debacles at Amphipolis (5.10.5) and Epipolae (7.44). He has Nicias complain about the lack of discipline of his Athenian troops, noting that they were 'by nature difficult to command' (7.14.2). Indeed, according to Xenophon, the Athenian levy took pride in treating its generals with disdain (*Mem.* 3.5.16; *Oec.* 21.4). Several extant Athenian court speeches attest to the problem of draft dodging and desertion (Lysias 14.5–15; 15.1; 16.15; Lycurgus 1.131, 147).[30] When the fourth-century general Phocion was challenged on his cautious policies, he replied that he would continue to advise the Athenians to avoid war until he could see 'the young men willing to hold their places in the formation, and the rich to pay taxes, and the politicians to stop embezzling public funds' (Plut. *Phoc.* 23.2) – implying that these basic requirements of discipline and integrity would never be met.

In addition to its lack of training and respect for military authority, the Athenian phalanx suffered from internal divisions caused by the physical and socio-economic inequalities inherent in a mass levy. Plato argues that having to fight side by side with the clumsy, pale, pudgy rich put thoughts of rebellion into the minds of the poor (*Rep.* 556c-e). At the same time, he notes that Socrates was resented by the other Athenians at Potidaea for his extreme endurance, which they took as a show of contempt for his fellow combatants (*Symp.* 220a-b), and that anyone who trained with weapons could expect to be mocked (*Lach.* 184c; *Leg.* 830d). Xenophon points out that many of the men in such militias were unfit for service due to their inexperience, old age, or poor physical condition (*Hell.* 6.1.5); Andocides complains that at Athens 'the old go on campaign while the young make speeches' (4.22). In Thucydides' enumeration of Athenian forces at the outset of the Peloponnesian War, just 45% of the hoplite levy is deemed fit to serve in the field army (2.13.6).

In short, while the Spartan phalanx is idealised as a uniquely cohesive and capable force, the Athenian hoplite body is represented as a rabble that shared none of its supposed strengths. Whether the difference between Athenian and Spartan hoplites was really so great in practice need not concern us here. The important point is that, for several Athenian authors, discussion of military matters involved endorsing Spartan or Spartan-inspired practices, sometimes at the explicit expense of Athenian ones. Whatever capabilities the Athenian hoplite possessed were clearly considered inadequate. Alternative methods were available that could improve his unit cohesion and effectiveness in battle.

Improving Unit Cohesion

The first of these methods was actually training the troops. It may be difficult for us to imagine that this was controversial, but in Thucydides we find Pericles declaring that the Athenians did not need rigorous exercise to fight well (2.39.1), and Plato presents an experienced general dismissing the concept altogether (*Lach.* 182d–184c). It was therefore necessary for Classical authors to make the case for military training repeatedly and at length.

Focusing on weapon proficiency, both Plato (*Rep.* 374d; *Leg.* 840a–c) and Aristotle (*Nic. Eth.* 1116b.7–8) made the point that the difference between trained warriors and amateurs was like that between craftsmen or athletes and ordinary people. Despite his worries that people might find his programme 'ridiculous', Plato outlined in detail how he believed a city-state's defenders ought to be prepared for their duties, through training with every known weapon, exercises in armour, and sham battles of various kinds (*Leg.* 794c, 813d–814b, 829a–831a, 833d-e). Xenophon clearly agreed; his description of Cyrus' efforts to turn his militia into a capable force includes several references to weapons training (*Cyr.* 2.1.21, 3.3.9, 3.3.50) as well as an account of a mock battle between groups of men (*Cyr.* 2.3.17–20).[31]

Crucially, in each of these passages, practice in the use of weapons is explicitly linked with morale, rather than being presented as a simple matter of 'honing personal skills'.[32] Mastery of their arms was supposed to fill the troops with faith in their own abilities and with contempt for the enemy. Thucydides already argued that if the men of Syracuse would begin to train for war, 'their valour would be led to surpass itself by the confidence which skill inspires' (Thuc. 6.72.4). Training would not merely make them stronger or faster – it would make them better men (Xen. *Mem.* 3.3.5–7). Once they had been prepared for battle in this way, they would be filled with a thirst for glory and a desire for victory that a general's speech alone could never produce (*Cyr.* 3.3.50–55). Xenophon's description of Agesilaus' efforts to improve his army in Asia Minor reads almost like a motivational movie script: after his inexperienced men suffer a minor defeat (*Hell.* 3.4.13–14), there is a flurry of exercise and competitive weapons training (16–17), followed by a second engagement (22–24) in which the self-assured expeditionary force functions like a well-oiled machine.

Some of the passages already cited suggest that Xenophon found formation drill even more important as a way to increase military effectiveness. On top of his suggestion that the Spartiates possessed unique powers of ad hoc unit formation thanks to their careful training, he stresses the imposing spectacle of an army marching in good order (*Oec.* 8.6–7) and the ease with which a well-drilled force could advance at speed while keeping its formation intact (*Cyr.* 3.3.57). This was not intended as mere fawning on Spartan methods. The example was meant to be followed. Xenophon uniquely describes the actual process of teaching untrained troops formation drill (*Cyr.* 2.2.6–10), and repeatedly points out the tactical abilities thus gained (*Hell.* 6.5.18–19; *Cyr.* 2.3.21–23, 3.3.70; *Lac.* 11.5–10). Plato, too, made tactical drill a part of his proposed training programme (*Leg.* 813e) – but Xenophon was much more assertive. Anticipating his readers' reluctance, he argued that 'none of this is difficult to learn' and that 'no one who can tell people apart can possibly get it wrong' (*Lac.* 11.6). In his view, given how easy it was to master Spartan drill, the Athenians did themselves a disservice by not adopting it immediately and reaping its obvious benefits.

Fundamental to this form of formation drill was a second method to improve the cohesion of the Athenian phalanx: the subdivision of the army into smaller units, each with its own commander, down to the level of the file. Thucydides stood in awe of the Spartan phalanx, consisting almost entirely of 'leaders leading leaders' (5.66.4) who maintained good order, deployed for battle quickly and carefully, and transmitted orders smoothly down the line. Xenophon described the organisation and use of the picked units of the Ten Thousand, subdivided according to the Spartan model, demonstrating that this system could easily be adopted by others (*An.* 3.4.21–23). He would presumably have given the Athenian hoplite body the same advice he gave to the cavalry:

> The city has divided the cavalry into ten separate *phylai*. I say that within these you should, first of all, appoint leaders of ten, after consulting each of the commanders of the *phylai*, choosing men in their prime who are ambitious to do some great deed. These should form the front rank. (…) The reasons why I like this formation are these. In the first place, all the men in the front rank are officers; and the obligation to do some great thing appeals more strongly to men when they are officers than when they are ordinary men. Secondly, when anything has to be done, the word of command is much more effective if it is passed to officers rather than to ordinary men.
>
> (Xen. *Eq. mag.* 2.2–6)

Xenophon confirms the tactical advantages noted by Thucydides – the effective transmission of orders, as well as the speed and precision with which a properly subdivided unit could be organised for battle (*Eq. mag.* 2.7–8; *Cyr.* 2.1.26). However, again, he adds an important moral dimension. Promotion to a leadership role was meant to invoke men's love of honour, inspiring them to give the right example to the rest of their file. Furthermore, in the process of organising their units, the officers got to choose whom to station directly behind them, and the men they picked got to choose the men behind them in turn, so that every man would trust the one who stood at his back. At the very rear, reliable men were to be posted, whose cheers were meant to encourage the whole line (Xen. *Eq. mag.* 2.4–5). Meanwhile, the formal grouping of men into distinct units was supposed to foster a sense of equality and camaraderie felt only by those who worked, ate and slept in each other's company (Xen. *Cyr.* 2.1.25, 28). To Xenophon, then, the creation of a hierarchy of sub-units was not only about enhancing order and tactical ability. It was about improving unit cohesion and combat effectiveness through the establishment of bonds of familiarity and trust within tightly knit groups of warriors led by ambitious junior officers. This seems to address very precisely the shortcomings of the Athenian phalanx outlined above.

A final method to increase the quality of this formation was the introduction of better discipline. Carelessness, lack of discipline and insubordination were known causes of defeat (Thuc. 6.72.3–4; Xen. *An.* 3.1.28; Plut. *Phoc.* 12.3, 26.1). Obedience made men reliable, which made it the essential precondition for all other reforms. As Xenophon's Socrates asked a newly appointed cavalry commander, 'Have you considered how to make the men obey you? Because without that, horses and horsemen, however good and gallant, are useless' (*Mem.* 3.3.8).[33] Getting the men to follow orders and face danger, in the belief that their commanders knew best, was expected to have two interconnected results. The first was practical: a general could not hope to carry out his battle plans unless his troops obeyed his orders. The second, again, was moral. Good discipline was supposed to make men industrious, loyal, and brave (Xen. *Ages.* 6.4; *Cyr.* 2.1.22; *Oec.* 21.5).

How could such discipline be obtained? Since the Spartans were perceived to enjoy the advantages of an obedient hoplite body, Athenian authors naturally looked to them for inspiration. Xenophon has one character wonder aloud when the Athenians will finally do the right thing and adopt the Spartan system of physical training (*Mem.* 3.5.15). Both Xenophon and Plato were advocates of the Spartan custom to inculcate strict obedience in citizens from an early age (Xen. *Lac.* 2.2–11; Pl. *Leg.* 808d–809a, 942c).[34] Plato, however, was willing to go much further:

> The main principle is this: that nobody, male or female, should ever be left without control, nor should anyone, whether at work or in child's play, grow habituated in mind to acting alone and on his own initiative, but he should live always, both in war and peace, with his eyes fixed constantly on his commander and following his lead; and he should be guided by him even in the smallest detail of his actions. (…) Anarchy must be utterly removed from the lives of all mankind, and of the beasts also that are subject to man.
>
> (*Leg.* 942a–d)

This chilling totalitarian vision stood in stark contrast to the Athenian democratic habit, maligned by philosophers and orators alike, to live exactly as they pleased (Thuc. 2.39.1; Lys. 14.11; Pl. *Rep.* 557b; Arist. *Pol.* 1317b.10–12). As noted above, the Athenians were also generally suspicious of military authority as such. Even Plato himself seems to have realised that these prevailing values rendered his vision unattainable. After his initial programmatic statement, the actual laws he prescribed to ensure good military conduct were no more strict than those of contemporary Athens (*Leg.* 943–945). Xenophon, too, understood the limits of harsh discipline; as a mercenary general, he had been put on trial by his soldiers for using his fists as a disciplinary measure (*An.* 5.8). To these authors, it would have seemed practically impossible to force Athenian hoplites to obey, and equally impossible to persuade them to make obedience a greater part of their civic ideal.

The only possibility was to secure obedience and good discipline by means other than force. In Xenophon's work, the responsibility falls entirely on the general. His monumental task was to make men *willing* to obey. The general's personal example was supposed to invoke the respect and admiration of the troops (*Ages.* 6.4; *Eq. mag.* 6.4–6; *Mem.* 3.3.9); a system of honours and rewards was supposed to encourage them to compete for his favour (*Cyr.* 1.6.20, 2.3.21–24, 8.1.2–4, 8.2.2–4; *Oec.* 5.15). In addition, Xenophon insists that any man who wishes to be a general had better be a good speaker, because he will be required to *persuade* his men that discipline and obedience are in their own interest (*Eq. mag.* 1.18–19, 1.22–24; *Mem.* 3.3.10–11; *Oec.* 13.9).[35] Only 'the superhuman and good and knowing leader' had any hope of inspiring eager obedience (*Oec.* 21.5–8), with all the advantages of cohesion and combat effectiveness that came with it.

Conclusion

The Athenian hoplite phalanx was a levy of eager amateurs. Its fighting ability rested on horizontal unit cohesion: it was wholly lacking in institutional organisation and training, but its members believed in their city's cause, and in combat they refused to let their friends and neighbours down. Their moral commitment gave their phalanx its strength. In a world of amateur armies, this was sometimes enough. Meanwhile, in a context of democratic and egalitarian ideology, tighter vertical unit cohesion and more extensive training were not welcome. The selfless bravery of the citizen hoplite was the military expression of the spirit of Athenian democracy, glorified as not merely sufficient, but ideal. This was the 'culture of combat' Crowley analysed in peerless fashion.

Athenian literary sources, however, reveal a range of critiques of this culture. These contemporary texts portray the Athenian hoplite body as ill-prepared, skittish, insubordinate, weak, and prone to fatal confusion. Their authors apparently regarded the moral pressure of the primary group and its values as an unsatisfactory foundation for the Athenian hoplite's performance in battle. While many Athenians, in the style of the French army before 1870, clung to the creed that 'we will always muddle through',[36] some recognised that even a few basic improvements in training, organisation, and discipline could raise the city's hoplites to a higher standard.

To the authors that advocated them, these reforms were much more than merely technical. Certainly, they served to improve the weapon proficiency and tactical ability of the hoplite body. But this was only one part of the story. Experienced commanders like Xenophon observed that armies with superior organisation and weapon skills derived unrivalled confidence from their abilities; that their unit hierarchies created strong ties between fellow combatants; that their discipline made them dependable in the face of danger. They connected the seemingly cold logic of formation drill and unit subdivision directly to morale and unit cohesion, justifying their proposed methods by stressing that they would not just improve the army, but each unit, each file, and each individual warrior. Far from offering mere supplementary skills to a hoplite's already existing will to fight, training and discipline were shown to be the potential cornerstones of that fighting spirit. In a strengthened framework of vertical cohesion, well-drilled warriors would not just fight more effectively, but also more bravely, trusting in their superiority and urging each other on. The trained, obedient hoplite was not just a superior fighter – he would outdo the amateur in moral commitment, too.

The recommendations of authors like Xenophon seem to have gone mostly unheeded. Perhaps the methods they proposed reeked too strongly of Sparta. Perhaps, as they themselves seem to have realised, their authoritarian ideas were simply incompatible with the values of freeborn Athenians. Even they could not have guessed the consequence: the rulers of Macedon adopted all the improvements cited here, and their professional armies brought the freedom of the Greeks to an end.

Notes

1 See for example Hanson (1989): 110, 117–25, (1995): 279–80; Lazenby (1991): 106–8; Hutchinson (2000): 190–1; Wheeler (2001): 174; van Wees (2004): 163; Eckstein (2005): 482–5; Lee (2006): 483; Chrissanthos (2013): 314.
2 Crowley (2012).
3 Crowley (2012): 3.
4 Crowley (2012) on training, 25–6, 81; on tactics, 2–3; on the absence of equivalents to platoon or section commanders, 124 (although he assumed their existence at 42–3); on pay, 107–9; on disciplinary measures, 106; on the lack of military units in the modern sense, 79. For an earlier litany of the Athenian hoplite's organisational shortcomings, see Ridley (1979): 513–17.
5 Though the concepts listed here are not part of Crowley's work, they allow for a neat summary of his argument in the terms of the current volume. I do not offer any in-depth exploration of these concepts here. For the essential theoretical framework, see the introduction to this volume. In the interest of clarity, I use these terms to denote ties of kinship, morality, and purpose (horizontal cohesion) on the one hand, and structures of organisation and command (vertical cohesion) on the other. I take horizontal cohesion to comprise both social and task cohesion; for the distinction, see MacCoun, Kier, and Belkin (2006).
6 The following sections revisit several points I explored in more detail in the context of a different argument: see Konijnendijk (2018): 39–71 and 139–53.
7 Anderson (1970): 84–91; Ridley (1979): 548; Osborne (1987): 145–6; Debidour (2002): 27–9. Many modern authors still subscribe to the view that hoplite combat required extensive training, and that the existence of such training can therefore be taken for granted; see for example, Rusch (2011): 14; Ober (2015): 31.
8 Crowley (2012): 81; see also, for example, Lazenby (1989): 69; Goldsworthy (1997): 8–10; van Wees (2004): 89–93; Rawlings (2007): 90; Lee (2013): 145–6; Chrissanthos (2013): 315–17.

9 Crowley (2012): 86–104.
10 The evidence for Greek military training is gathered in Konijnendijk (2018): 42–70; the most comprehensive earlier treatments are Anderson (1970): 84–110 and Ridley (1979): 530–48. The specific recommendations of Classical authors will be discussed in more detail below.
11 Crowley (2012): 42–3; see also Ducrey (1985): 69–72; Lee (2006): 483; Matthew (2012): 171–2.
12 The Continental Army of the American War of Independence offers an insightful parallel. With this army of volunteers, according to one contemporary observer, 'it was almost impossible to advance or retire in the presence of an enemy without disordering the line or falling into confusion'. It took a Prussian officer four days to teach basic formation drill to George Washington's personal guard, which was then required to pass on his lessons to the rest of the army. See Royster (1979): 218; Lockhart (2008): 90–3. I am grateful to Hunter Higgison for bringing these works to my attention.
13 Lazenby (1991): 89; Lee (2004): 289–90, 302–3; van Wees (2004): 99–100; Lendon (2005): 74–5; Hunt (2008): 129–30.
14 Before the final battle in the harbour at Syracuse, Thucydides (7.61.3) has Nicias plead specifically to those 'who already have much experience of war' to show courage and give the others the right example.
15 Lazenby (1991): 90; Krentz (1991): 115–16. Even among Spartan-led troops, confusion could ensue if parts of the line were not aware that the advance had begun: see Xen. *Hell.* 6.4.13.
16 The *salpinx* was also used to give certain signals off the battlefield; the different context would presumably have made the different intention of the signal clear to the men. On the roles of the *salpinx* in Classical warfare, see Krentz (1991).
17 Crowley (2012): 70–9.
18 Crowley (2012): 124.
19 *Contra* Hutchinson (2000): 190–1; Crowley (2012): 124–5.
20 Ridley (1979): 513–14; Hamel (1998): 59–63; Hornblower (2000): 57–61, 72–3; van Wees (2004): 108–12; Eckstein (2005): 483–4; Lendon (2005): 74–7; Christ (2006): 40–1; Rawlings (2013): 13, 20–1; Chrissanthos (2013): 315–17.
21 Crowley (2012): 106.
22 Bettalli (2002): 117–19; Hornblower (2004): 247, 263; Harris (2010): 414–15, (2015): 84–8.
23 Hamel (1998): 62, 118–21; the excursus at 140–57 identifies no fewer than 65 possible cases of Athenian generals put on trial. See also Hornblower (2000) for further examples of Greek warriors disciplining their own commanders through litigation or outright violence.
24 The same contrast between appropriate (Spartan) and insufficient (Athenian) legal consequences of cowardice was drawn by the orator Lycurgus some decades later (Lycurg. 1.129–130).
25 Eckstein (2005): 485. Note Crowley's (2012): 43–6 discussion of the strong bonds within the deme levies that formed the building blocks of the Athenian phalanx.
26 For a detailed treatment of this idea and its relation to the Theban Sacred Band, see Leitao (2002).
27 Crowley (2012): 80–104 has argued that Athenian hoplites' willingness to do violence can be taken for granted, and this would have had a positive effect on their performance in battle. In addition, though he does not address this directly, it may be assumed that their commitment to common goals (task cohesion) was typically very high; it is frequently invoked in pre-battle speeches (Thuc. 4.95, 6.68.3, 7.61.1; Xen. *Hell.* 2.4.13–14, 17) and even receives general praise from contemporary commentators on Athenian democracy (Hdt. 5.78; Thuc. 2.39.1–2). However,

Classical authors persist in their criticism of the Athenian militia regardless. The suggestions for improvement they offer address different forms and aspects of unit cohesion. Therefore, the acknowledged martial values of hoplites and the strength of the ties to their cause does not affect the argument given here, and will not be discussed in detail.

28 See, for example, Hom. *Il.* 12.265–276 and 17.414–422.

29 The echoes of Spartan practice in the fictional army of Xenophon's *Education of Cyrus* are very clear; see for example Christesen (2006): 52–3.

30 Later sources offer examples of Athenian (or partly Athenian) forces drained by desertion, such as Phocion's army in Euboea in the 340s (Plut. *Phoc.* 12, where Athenian troops are described as 'disorderly and chattering and worthless') and the combined Greek army at Krannon in 322 (Diod. 18.17.1; Plut. *Phoc.* 26.1).

31 Xenophon also recommends sham battles to train cavalry (*Eq. mag.* 3.11–13).

32 This is how Hutchinson (2000): 61, characterised the training of the Ten Thousand, which must have been part of the inspiration for Xenophon's fictional account.

33 Xenophon repeats the same sentiment throughout his body of work: see *Eq. mag.* 1.7; *Cyr.* 1.6.13–14, 8.1.2; *Mem.* 3.4.8.

34 This was sometimes represented as the traditional way of the Athenians of old, too; see Ar. *Nub.* 961–983; Xen. *Mem.* 3.5.14.

35 Christ (2006): 42–3; Chrissanthos (2013): 317.

36 'On se débrouillera toujours': cited in Wawro (2000): 108.

Bibliography

Anderson, J. K. 1970: *Military Theory and Practice in the Age of Xenophon*, Berkeley.

Bettalli, M. 2002: 'La disciplina negli eserciti delle *poleis:* il caso di Atene'. In M. Sordi (ed.), *Guerra e Diritto nel Mondo Greco e Romano*, Milan: 107–121.

Chrissanthos, S. G. 2013: 'Keeping military discipline'. In B. Campbell and L. A. Tritle (eds.), *The Oxford Handbook of Warfare in the Classical World*, Oxford: 312–329.

Christ, M. R. 2006: *The Bad Citizen in Classical Athens*, Cambridge.

Christesen, P. 2006: 'Xenophon's *Cyropaedia* and military reform in Sparta', *JHS* 126: 47–65.

Crowley, J. 2012: *The Psychology of the Athenian Hoplite: The Culture of Combat in Classical Athens*, Cambridge.

Debidour, M. 2002: *Les Grecs et la Guerre, Ve–IVe Siècles: De la Guerre Rituelle à la Guerre Totale*, Monaco.

Ducrey, P. 1985: *Guerre et Guerrier dans la Grèce Antique*, Paris.

Eckstein, A. M. 2005: 'Bellicosity and anarchy: Soldiers, warriors, and combat in Antiquity', *The International History Review* 27 (3): 481–497.

Goldsworthy, A. K. 1997: 'The othismos, myths and heresies: The nature of hoplite battle', *War in History* 4 (1): 1–26.

Hamel, D. 1998: *Athenian Generals: Military Authority in the Classical Period*, Leiden.

Hanson, V. D. 1989: *The Western Way of War: Infantry Battle in Classical Greece*, New York.

Hanson, V. D. 1995: *The Other Greeks: The Family Farm and the Agrarian Roots of Western Civilization*, New York.

Hanson, V. D. 2000: 'Hoplite battle as ancient Greek warfare: When, where, and why?'. In H. van Wees (ed.), *War and Violence in Ancient Greece*, London: 201–232.

Harris, E. 2010: 'The rule of law and military organisation in the Greek p*olis*'. In G. Thür (ed.), *Symposion 2009: Vorträge zur griechischen und hellenistischen Rechtsgeschichte*, Vienna: 405–417.

Harris, E. 2015: 'Military organisation and one-man rule in the Greek *polis*', *Ktema* 40: 83–91.

Hornblower, S. 2000: 'Sticks, stones and Spartans: The sociology of Spartan violence'. In H. van Wees (ed.), *War and Violence in Ancient Greece*, London: 57–82.

Hornblower, S. 2004: '"This was decided" (*edoxe tauta*): The army as p*olis* in Xenophon's *Anabasis* – and elsewhere'. In R. Lane Fox (ed.), *The Long March: Xenophon and the Ten Thousand*, New Haven: 243–263.

Hunt, P. 2008: 'Military forces'. In P. Sabin, H. van Wees and M. Whitby (eds.), *The Cambridge History of Greek and Roman Warfare, Volume 1: Greece, the Hellenistic World and the Rise of Rome*, Cambridge: 108–146.

Hutchinson, G. 2000: *Xenophon and the Art of Command*, London.

Konijnendijk, R. 2018: *Classical Greek Tactics: A Cultural History*, Leiden.

Krentz, P. 1991: 'The *salpinx* in Greek warfare'. In V. D. Hanson (ed.), *Hoplites: The Classical Greek Battle Experience*, New York: 110–120.

Lazenby, J. 1989: 'Hoplite warfare'. In J. Hackett (ed.), *Warfare in the Ancient World*, London: 54–81.

Lazenby, J. (1991). '*The Killing Zone*'. In Hanson, V. D. (ed.), Hoplites: The Classical Greek Battle Experience, New York: 87–109.

Lee, J. W. I. 2004: 'The *lochos* in Xenophon's *Anabasis*'. In C. Tuplin (ed.), *Xenophon and His World*, Stuttgart: 289–317.

Lee, J. W. I. 2006: 'Warfare in the Classical age'. In K. H. Kinzl (ed.), *A Companion to the Classical Greek World*, Malden, MA: 480–508.

Lee, J. W. I. 2013: 'The Classical Greek experience'. In B. Campbell and L. A. Tritle (eds.), *The Oxford Handbook of Warfare in the Classical World*, Oxford: 143–161.

Leitao, D. 2002: 'The legend of the Sacred Band'. In M. C. Nussbaum and J. Sihvola (eds.), *The Sleep of Reason: Erotic Experience and Sexual Ethics in Ancient Greece and Rome*, Chicago: 143–169.

Lendon, J. E. 2005: *Soldiers and Ghosts: A History of Battle in Classical Antiquity*, New Haven.

Lockhart, P. 2008: *The Drillmaster of Valley Forge: The Baron de Steuben and the Making of the American Army*, New York.

MacCoun, R. J., Kier, E. and A. Belkin, 2006: 'Does Social Cohesion Determine Motivation in Combat? An Old Question with an Old Answer', *Armed Forces & Society* 32 (4): 646–654.

Matthew, C. A. 2012: *A Storm of Spears: Understanding the Greek Hoplite at War*, Barnsley.

Ober, J. 2015: *The Rise and Fall of Classical Greece*, Princeton.

Osborne, R. 1987: *Classical Landscape with Figures: The Ancient Greek City and its Countryside*, London.

Rawlings, L. 2007: *The Ancient Greeks at War*, Manchester.

Rawlings, L. 2013: 'War and Warfare in Ancient Greece'. In B. Campbell and L. A. Tritle (eds.), *The Oxford Handbook of Warfare in the Classical World*, Oxford: 3–28.

Ridley, R. T. 1979: 'The hoplite as citizen: Athenian military institutions in their social context', *Antiquité Classique* 48: 508–548.

Royster, C. 1979: *A Revolutionary People at War: The Continental Army and American Character, 1775–1783*, Chapel Hill.
Rusch, S. M. 2011: *Sparta at War: Strategy, Tactics, and Campaigns, 550–362 BC*, London.
Van Wees, H. 2004: *Greek Warfare: Myths and Realities*, London.
Wawro, G. 2000: *Warfare and Society in Europe, 1792–1914*, Abingdon.
Wheeler, E. L. 2001: 'Firepower: Missile weapons and the "Face of Battle"', *Electrum* 5: 169–184.

2 The Rhodian Slingers in Xenophon's *Anabasis*

C. W. Marshall

The military virtues of the sling within the context of organized classical warfare can helpfully be framed with reference to the formation of a unit of Rhodian slingers in Xenophon's *Anabasis* 3.3. Because the company is created mid-campaign, it is possible to isolate specific military needs faced by the Ten Thousand that required the reassignment of Rhodians to a specialized light-armed force. These circumstances demonstrate clearly the tactical value of light-armed troops (*psiloi*, a term that comprises slingers, archers, and javelin throwers), and this corroborates the specific military virtues of the sling that Xenophon presumes in his discussion.[1] More importantly, Xenophon's account authorizes conclusions about the effectiveness of an impromptu light-armed unit.

The slinger unit is formed following the initial struggles faced by the Greek mercenaries following the Battle of Cunaxa (401 BCE). Rhodians are removed from their existing units and reconstituted because they possess a technology absent from the opposing Persian forces (and, implicitly, absent from the rest of the Greek forces):

> [3.3.16] ἀκούω δ' εἶναι ἐν τῷ στρατεύματι ἡμῶν Ῥοδίους, ὧν τοὺς πολλούς φασιν ἐπίστασθαι σφενδονᾶν, καὶ τὸ βέλος αὐτῶν καὶ διπλάσιον φέρεσθαι τῶν Περσικῶν σφενδονῶν. [17] ἐκεῖναι γὰρ διὰ τὸ χειροπληθέσι τοῖς λίθοις σφενδονᾶν ἐπὶ βραχὺ ἐξικνοῦνται, οἱ δὲ Ῥόδιοι καὶ ταῖς μολυβδίσιν ἐπίστανται χρῆσθαι.
>
> [3.3.16] Now I am told that there are Rhodians in our army, that most of them understand the use of the sling, and that their missile carries no less than twice as far as those from the Persian slings. [17] For the latter have only a short range because the stones that are used in them are as large as the hand can hold; the Rhodians, however, are versed also in the art of slinging leaden bullets.

The company is small (200 soldiers according to *An.* 3.3.20), but will prove effective at combating the skirmishers attacking the Ten Thousand. Rhodian slingers use lead bullets, which are capable of being effective at

DOI: 10.4324/9781315171753-3

twice the Persian effective range when using oversized stones, which were negatively affected by a slower speed of release and then a greater impact of wind resistance.[2] Xenophon's account of the Rhodian unit is the most detailed account of the sling's use in warfare to survive from the classical period.[3]

The following discussion examines the company's effectiveness along three axes: technology, cohesion, and tactics. The unit's cohesion is integral to its successful deployment, but this cannot be separated from technological and tactical considerations. In part because of the nature of the surviving evidence, and in part because of concerns specific to the role of light-armed troops in combat, the discussion of cohesion in this chapter possesses different emphases than many of the other examples discussed in this volume. It would go beyond the evidence available to speculate too far, but it is possible to gain a deeper appreciation of how slingers and this company in particular operated as a discrete unit within a larger campaign. The Rhodian slingers in Xenophon's *Anabasis* become a valuable reference point for understanding the limits of unit cohesion in the ancient world.

The Sling as a Technology

Slings were employed from pre–Bronze Age Mesopotamia to the medieval period, and they continued to prove effective despite developments in armour and tactics. Early caches of sling bullets from Mesopotamia (c. 6000 BCE) and Syria (c. 3500 BCE) show the antiquity of slings in (proto-)urban defence.[4] The sling appears prominently in the early Iron Age story of David's victory over the Philistine Goliath (1 Samuel 17).[5] As a skirmishing weapon, the sling was in use throughout the classical period (e.g. Herodotus 7.158.4 at Syracuse in 480 BCE; Thucydides 6.22.1, 6.25.2, 6.43 at Sicily in 415); Xenophon *Hell.* 4.2.16 at Sparta in 394). Slingers appear to have been crucial in the attack on Olynthus in 348.[6] Slingers served as auxiliary troops in the Roman army (*funditores*) as new technologies (such as the staff-sling and dart sling) are introduced,[7] and into medieval Europe.[8]

For Xenophon, weapon range is the primary virtue of the sling. His account has appeared counter-intuitive to some readers, and so requires examination because it demonstrates his precise understanding of the weapon's capabilities. The Persian archers and slingers possessed a greater effective range than the Cretan archers and the Greek javelins:

> [3.3.15] νῦν γὰρ οἱ πολέμιοι τοξεύουσι καὶ σφενδονῶσιν ὅσον οὔτε οἱ Κρῆτες ἀντιτοξεύειν δύνανται οὔτε οἱ ἐκ χειρὸς βάλλοντες ἐξικνεῖσθαι.
>
> [3.3.15] For at present the enemy can shoot arrows and sling stones so far that neither our Cretan bowmen nor our javelin-men can reach them in reply.

Effective range is relevant because the deployment of the Cretan archers among the Ten Thousand was suboptimal, as Xenophon has explained:

> [3.3.7] οἱ δὲ ὀπισθοφύλακες τῶν Ἑλλήνων ἔπασχον μὲν κακῶς, ἀντεποίουν δ' οὐδέν· οἵ τε γὰρ Κρῆτες βραχύτερα τῶν Περσῶν ἐτόξευον καὶ ἅμα ψιλοὶ ὄντες εἴσω τῶν ὅπλων κατεκέκλειντο, οἱ δὲ ἀκοντισταὶ βραχύτερα ἠκόντιζον ἢ ὡς ἐξικνεῖσθαι τῶν σφενδονητῶν.
>
> [3.3.7] And the Greek rearguard, while suffering severely, could not retaliate at all; for the Cretan bowmen not only had a shorter range than the Persians, but besides, since they had no armour, they were shut in within the lines of the hoplites, and the Greek javelin could not throw far enough to reach the enemy's slingers.

Since there were only 200 Cretan archers (*An.* 1.2.9) and they were unarmoured,[9] they had been placed in a defensive position, which moved them further back from the line of battle.[10] This deployment, within the rectangle formed by the hoplites, reduced the archers' ability to hit distant targets and required that the archers shoot over the heads of their front-line comrades. Firing in an arc in this way can potentially increase absolute range for a bow (this is the technique described for the Persians by Herodotus 7.226), but to be most effective requires a greater number of archers than the Ten Thousand possessed.

Anabasis 3.4.17 provides one reason the slings were more effective than bows: μεγάλα δὲ καὶ τὰ τόξα τὰ Περσικά ἐστιν· ὥστε χρήσιμα ἦν ὁπόσα ἁλίσκοιτο τῶν τοξευμάτων τοῖς Κρησί, καὶ διετέλουν χρώμενοι τοῖς τῶν πολεμίων τοξεύμασι, καὶ ἐμελέτων τοξεύειν ἄνω ἱέντες μακράν ('The Persian bows are also large, and consequently the Cretans could make good use of all the arrows that fell into their hands; in fact, they were continually using the enemy's arrows, and practiced themselves in long-range work by shooting them into the air'). The training regimen, increasing range by shooting in parabolic arcs, suggests that the Cretan archers were not used to firing for distance. This training can be construed as an effort to reinforce cohesion in the light of a new tactical and logistical situation. Technological limitations also impacted the archers: a smaller bow means a smaller draw and consequently less force transferred to the projectile, and reusing enemy arrows is necessarily suboptimal, especially when they are oversized (and so longer than those for which the Cretan bows were designed).[11] Given the Greek isolation in Persian territory, where they were unable to purchase or easily make arrows of an appropriate size for their own bows, the re-use of oversized Persian arrows, whether captured in battle or otherwise obtained, is economical but mechanically inefficient.

The context surrounding this analysis (3.4.15–17) clarifies Xenophon's understanding of the technological benefits of the sling among the *psiloi.*

Xenophon is unambiguous that Rhodian slingers had a greater effective range than both the Persian bowmen and the Persian slingers: καὶ οὐκέτι ἐσίνοντο οἱ βάρβαροι τῇ τότε ἀκροβολίσει· μακρότερον γὰρ οἵ γε Ῥόδιοι τῶν Περσῶν ἐσφενδόνων καὶ τῶν τοξοτῶν (3.4.16: 'and the barbarians were no longer able to do any harm by their skirmishing at long range; for the Rhodian slingers carried farther with their missiles than the Persians, farther even than the Persian bowmen').[12] Given this, the formation of a troop of slingers seemed prudent: the Greek slingers were capable of greater range than other missile troops.[13]

This same passage addresses the issue of resupply, which would be essential for the unit to have any long-term effectiveness. Slings could be woven from a variety of natural fibres that were easily available as the Greeks travelled: ηὑρίσκετο δὲ καὶ νεῦρα πολλὰ ἐν ταῖς κώμαις καὶ μόλυβδος, ὥστε χρῆσθαι εἰς τὰς σφενδόνας (3.4.17: 'In the villages, furthermore, the Greeks found gut [sic] in abundance and lead for the use of their slingers').[14] When he formed the company, Xenophon had suggested buying any slings that already existed (brought by mercenaries, perhaps for hunting), 'and likewise pay anyone who is willing to plait new ones' (3.3.18 τῷ δὲ ἄλλας πλέκειν ἐθέλοντι ἄλλο ἀργύριον τελῶμεν).[15] The sling is light, portable, and inexpensive (Vegetius 1.16). Slings can be used as belts or straps; they add no encumbrance, and can be manufactured by hand without special techniques. They can also be repurposed, as at *Iliad* 13.598–600 when one is used as a bandage. Sling bullets can also be produced without appreciable cost. Appropriate water-rounded stones can literally be picked up by the side of a river (1 Sam. 17:40, Livy 38.29.4), and lead bullets can be manufactured without specialist technological requirements.[16] Whatever material was employed for the slings, it was readily available in villages but not among the soldiers, as was the lead that could be used for sling bullets.[17]

The sling is also versatile in the possibilities for its deployment. Lee rightly emphasizes the accuracy of the sling at short range, and how the sling could be effective in close-quarters and in urban combat, and even in combat indoors.[18] Anecdotes about the accuracy of trained slingers isolate extraordinary incidents (Judges 20:16, Livy 38.29.7, Vegetius 1.16 and, against mahouts on invading elephants, 3.24), and demonstrate the widespread understanding that bullets could be aimed precisely. It is not possible to identify a preferred technique for ancient slingers, even presuming one existed. Advocates of a single overhand throw and a single underarm swing both exist, and no doubt a range of techniques could prove effective (Vegetius 2.23).[19] Assyrian reliefs positioning slingers behind archers not only points to their greater range, but also to their use for bombardment, as overhand slinging can cause bullets to arc high over the heads of those in front.[20] Slingers could operate effectively with shields (Vegetius 3.14, and possibly Tyrtaeus fr. 11.35–38) and could even be deployed as missile troops on ships (Thuc. 7.70.5; Florus 1.43.8).

Unit Cohesion

I have taken time to outline what Xenophon says about the sling explicitly, and to situate it within the wider context of the sling's use in the ancient world, since the effectiveness of the sling as a military technology is not always recognized despite its perseverance over time. Within the context of the theme of this volume, two details are crucial. First, circumstances emerged during the prolonged march of the Ten Thousand in which it proved advantageous to redeploy some soldiers as slingers, separating them from their assigned duties and forming a Rhodian unit. Rhodian slingers had a favourable reputation that Xenophon as an Athenian would appreciate (Thucydides 6.43.1 identifies a unit of Rhodians serving with the Athenian troops in Sicily), but there is no evidence of Rhodians being recognized as slingers before this by other *poleis*, and no particular reason Rhodians would be better slingers than anyone else.[21] Rather, Xenophon capitalized on an opportunity to increase the effective range of his missile troops with the formation of a new unit on campaign. Secondly, existing missile troops were also facing issues of supply and the army was constantly on the move, over new and unfamiliar terrain. The introduction of slingers, the majority of which were united by a shared ethnic identity on Rhodes, allowed the improvised unit to benefit immediately from a kind of intimacy or at least fellow-feeling amongst its members.

Social cohesion among soldiers is an essential part of combat readiness and performance, but considerable evidence has shown that its impact is not independent of other factors, but 'are the products of an array of inputs, policies, processes, and intangible factors in addition to unit cohesion, including leadership, training, mission, equipment, and logistical support, as well as weather, terrain, and enemy characteristics'.[22] Xenophon does not address all of these factors, and his text cannot be pressed to provide them. Nevertheless, this interdependence of variables is usefully examined in the light of the formation of the slinger unit. Scholarship that examines cohesion in terms of vertical cohesion and horizontal cohesion, with the latter further considered in terms of social cohesion and task cohesion, provides a substantial foundation for framing the discussion.[23] The broad categories of horizontal cohesion can be distinguished as follows:

> Social cohesion refers to the nature and quality of the emotional bonds of friendship, liking, caring, and closeness among group members. … Task cohesion refers to the shared commitment among members to achieving a goal that requires the collective efforts of the group.[24]

Vertical cohesion describes the relationship between leaders and followers (and is also known as rank cohesion).[25]

For the original mission of the mercenary army in the *Anabasis*, there is reason to believe that all of these factors are to some degree lacking. The mercenaries in 401 BCE might be thought to lack social cohesion, since they come from various *poleis* in Greece, which has just seen the end of the Peloponnesian War.[26] They might be thought to lack task cohesion since they are not invested in their mission apart from their pay. Finally, they might be thought to lack vertical cohesion, since their march against Artaxerxes II is not explained to them, and their commander, Cyrus the Younger, has not revealed his true purpose. This changes after the Battle of Cunaxa, and in *Anabasis* 3 and 4 factors associated with effective cohesion begin to emerge. Xenophon is one of five leaders elected by the soldiers (*Anab.* 3.1.26, 47), and it is possible to see his efforts to motivate and his responsiveness to soldiers' concerns in his anecdotes (e.g. *Anab.* 3.2.9, where Xenophon proposes a sacrifice that wins united approval from the soldiers, and 3.4.41–49, where his presentation of personal physical hardship might be thought to add veracity to an otherwise self-indulgent account). This change in leadership establishes improved vertical cohesion, and the shared mission to return to Greece offers a clear and achievable task to unite the mercenaries.[27] Even this might be thought to be rather imprecise, since '[t]o build task cohesion, it is not sufficient to emphasize the importance of the team's goals; units need their leaders to help them understand how to achieve those goals'.[28] For the Ten Thousand, the strength of any horizontal cohesion remained largely unaffected.

The Rhodian unit provides a good test case for understanding ancient military cohesion, even though Xenophon's account does not answer all questions a modern scholar may ask. The unit exhibits characteristics of social cohesion (a shared Rhodian identity) and more focused task cohesion (protecting the rearguard from skirmishers), as well as vertical cohesion (Xenophon earning trust through the redeployment of volunteer specialists). The initial composition of the Ten Thousand had not included a unit of slingers and Rhodians in the hoplite army were not serving together originally, in the way the Cretan archers were. In forming a new unit, Xenophon identified a dispersed population and provided a unifying identity for them, in order to counter specific strategic issues that faced the army on its return journey. Significantly, the unit was formed of volunteers, who were further incentivized with some relief from duties (3.3.18 ἀτέλεια). This suggests that not all of the unit's members needed to be Rhodian, and possibly not all Rhodians were included. Some heterogeneity will have existed within the new unit. Nevertheless, by identifying the unit as 'Rhodian', Xenophon creates a presumptive cultural uniformity, and affinity will already exist among the individual soldiers.[29]

Establishing a shared identity within the new contingent was particularly urgent, given that the slingers were to be variously stationed across the entire line of the rearguard:[30] ἐπεὶ δὲ διαταχθέντες οἱ Ῥόδιοι ἐσφενδόνησαν καὶ οἱ [Σκύθαι] τοξόται ἐτόξευσαν καὶ οὐδεὶς ἡμάρτανεν ἀνδρός (*Anab.* 3.4.15: 'But when the Rhodian slingers and the bowmen, posted at intervals here and there, sent back an answering volley, and not a man among them missed his mark').[31] One way to understand this deployment, if the Rhodians had been drawn from amongst the hoplites, is that slingers continued to be stationed with their previous units, but were now serving as missile troops instead of hoplites.[32] Such a deployment would allow individuals additionally to maintain cohesion and fellow-feeling with their established messmates while still introducing a new technology to the armament of the Ten Thousand, and providing them with a specialist identity that created additional horizontal bonds. The re-distribution of Rhodians need not therefore have left holes among the established soldiery, requiring further adjustments of various kinds all along the line. This understanding of διαταχθέντες allows individuals redeployed to a new specialist unit of predominantly Rhodian slingers to create a new unit identity even if they were not marching side by side through the day, and at the same time continue to enjoy the bonds of fellowship they have already established. Modern studies demonstrate that soldiers believe that social cohesion is an important motivation in combat, even if evidence does not support that belief.[33] Horizontal cohesion allowed the slingers to operate effectively in their new circumstances.

The cohesion that existed among the Rhodian slingers that Xenophon describes is not primarily of tactical benefit, but rather it was something that emerged in social and cultural bonds that developed on the march. The introduction of a new technology to the line of battle, as discussed in the previous section of this chapter, demonstrated a flexibility that depended on successful vertical cohesion already being in place. The new unit may have fought as individuals, but were given a collective identity predicated on shared ethnic and cultural values that would allow deep reinforcement as the march continued. When called upon to fight, the soldiers possessed social bonds that had developed outside of combat that allowed them to fight more effectively as skirmishers, as will be discussed in the next section of this chapter.

Xenophon had found a way to preserve morale and combat effectiveness despite the redeployment. This success allowed the slingers to operate effectively in new tactical circumstances and demonstrates the creation of meaningful cohesion within this company. While the disposition of slingers across the rearguard might generate a dual allegiance, to the new Rhodian unit and (perhaps) to their previous assignment as well, for the majority of the army continuity with previous practice would be maintained. Unlike the Cretan archers, who were established as an identified military unit previously in the campaign (*Anab.* 1.2.9) along with the Greek javelins, the Rhodians lacked cohesion from long-standing familiarity with each other. Xenophon's language suggests the impromptu group was operating under shared orders, with shared responsibilities; his regular reference to them as Rhodian, to be associated with the Cretan archers, created a sense of shared ethnic identity regardless of the actual ethnic composition of the volunteers; and their deployment against Persian skirmishers helped unify them with a precise mission. These three factors (which map precisely onto vertical cohesion, social cohesion, and task cohesion) demonstrate that Xenophon treated them as a distinct combat unit meant to function as a cohesive body. The previous existence of the Cretan archers (also composed of 200 *psiloi*) could additionally have created a kind of inter-unit cohesion, making the combined light troops more effective by increasing their number and their effective range.

Though a convenient shorthand, it is not certain that all the volunteer slingers shared a Rhodian heritage. While 'task cohesion was the most important determinant of group performance',[34] the perceived social connection derived from the shared presumptive place of origin could accomplish much: perception of social cohesion was as important as the reality. This might be particularly true since the very notion of what it meant to be Rhodian was itself in transition in 401. Rhodes had united into a synoecism in 408/7 following its revolt from Athens, and the oligarchic government now seemed focused on naval combat, providing ships for the Spartan fleet (Xen. *Hell.* 2.1.15, 17). The unified Rhodes additionally controlled territories on the mainland (the Rhodian *peraea* in southwestern Asia Minor),[35] and all these inhabitants would still be negotiating what constituted Rhodian identity. Mercenaries could therefore identify as Rhodian without it needing to convey a specific ethos or even political allegiance. Their identity as slingers was sufficient to provide unit cohesion.

Skirmishing

Greek phalanx warfare forced soldiers to kill at close range. Hanson ties this to the disdain traditionally felt for ranged weapons, especially the bow.[36] While this sentiment is not as unproblematic as is often suggested, the psychological difficulty of killing other humans at close range has been demonstrated in modern times across cultures, and would apply in antiquity as well.[37] In the phalanx, unit cohesion is reinforced by the physical dependence of each soldier in the front line on the shield of the man to his right for protection.[38] Xenophon's decision to interrupt this line with ranged troops reflects an understanding of his particular circumstance: because the Persians they were fighting were primarily skirmishers, the presence of 'soft' targets in the line were strategically acceptable, where they would not have been in straightforward hoplite warfare when topography was not an issue.[39] On the march in unknown, uneven terrain, missile troops were increasingly effective, and could overcome hoplites.[40] Xenophon's formation of the sling unit can be understood to emerge because of geographical features of the terrain through which the Greeks travel and the resulting nature of skirmish-based conflict: Xenophon's tactical deployment of the slingers is landscape-responsive.

Skirmishing is a form of ranged combat with its own fluid dynamic. Xenophon described Cretan bows as smaller than Persian and forcing the use of oversized arrows (3.4.17). Consequently, Persian missile troops could fire further than Cretan archers (3.3.15). Nevertheless, at *Anabasis* 3.4.15, both slings and bows are able to hit Persian missile troops. While this is not a concern for the slingers (who possess twice the range of the Persians, 3.3.17), the range of the Cretan archers appears inconsistent. There are two ways of reconciling this difficulty without resorting to emending the text. First, it is possible that the Cretans had kept a reservoir of their own arrows, properly fitted for their bows, which they could use when full range was needed. This explanation recognizes the scarcity of supplies that the Greek mercenaries faced and would explain why Xenophon spells out the difference in bow size (3.4.17). It is preferable, however, simply to recognize that light-armed troops in battle do not simply stand still when firing at one another. Movement is constant, as missile troops approach their enemy and reach what they feel will be a position from which they can hit their foe and fire. There is no magic distance for range, standing at which one can always achieve a hit. It is always better to be closer, and the reality of warfare for light-armed troops will have required constantly negotiating those optimal distances. This is easily forgotten today. Many scholars discuss the errors and difficulties of modern experimentation, as they try to assign absolute values to the range of missile weapons.[41] This is a mug's game, and there is too much incentive for scholars to overestimate combat distances when thinking in these terms. Soldiers have targets and need to be at a range at which they can reliably hit, and that will often fall short of their maximum distance. Men move into and out of effective range as they seek targets or avoid becoming them, and in this fluid context the Rhodians have two advantages: their greater range keeps them safer, and the trajectory and small size of their lead bullets makes it harder for an enemy to see and avoid incoming missiles. This is the reality being reflected in *Anabasis* 3.4.15–17.

The sling could also be used for bombardment, raining bullets from on high to break up a military formation.[42] A missile barrage hitting an enemy line can have a number of effects: 'This barrage caused confusion and demoralized the enemy … This helped break up the advancing force into a disorganized charge, that the solid line of defenders could more readily defeat'.[43] When used for bombardment, light troops could maintain a distance between hoplites and the attackers, creating a defensive no-man's-land for protection and degrading target cohesion.[44] The additional range provided by slings would necessarily prove effective. Offensively, skirmishers can test a phalanx's integrity (Thuc. 4.33, 5.10; Diod. 15.32.4). Xenophon describes how enemy sling bullets hit his soldiers' shields and resound (*Anab.* 4.3.29).[45] This could offer a psychological dimension to the use of slings in warfare. The existence of a lead bullet with the name of the Lydian satrap Tissaphernes suggests that this dimension was present at the time of the Ten Thousand.[46] The principal benefit of the Rhodian slingers among the Ten Thousand was their increased range, providing the attackers with 'strategic immunity'.[47]

Sling bullets offer a narrower lateral profile than arrows, and consequently are less subject to wind than arrows and so preserve their momentum over a greater distance.[48] There is a smaller reduction of force over long range than with other missiles, and bombardment additionally adds the force of gravity. The oblong, biconical shape of sling bullets increases the likelihood of penetration, which is governed by the surface area of impact. The Hippocratic text *On Head Wounds*, which is roughly contemporary with Xenophon, distinguishes cranial injury caused by missiles (where again, the bombardment from sling bullets would be the most likely circumstance for head injuries specifically; *On Head Wounds* 11.3). Celsus (*On Medicine* 7.5.4, first century CE) discusses techniques of removing shot from bone joints and deep injuries to soft tissue.[49] There is a tactical advantage in causing injury to an enemy soldier, since his care will occupy other soldiers as well: a wound might remove a greater number of combatants than a kill, reducing the integrity and cohesion of any enemy formation.

Enemy skirmishers needed to expose themselves to sling attack merely to get into the extreme range for their own missile weapons. Even if the Greek sling bullets were not proving deadly, they could be a consistent enough annoyance that they interfered with the established routines that the Persians had enjoyed in maintaining their own cohesion. An occasional sling bullet could hurt or wound, and the maintenance of any military formation for the attacking skirmishers would have become more difficult, with consequent demoralizing effects on the Persians. Xenophon does not focus on this dimension in his account, but he does emphasize the repeated successes of the rearguard after his strategic innovation. For a minimal investment, slings can disrupt a battle formation, maintain a defensive perimeter, and inculcate fear into the enemy (cf. Vegetius 1.16).

Thinking in terms of casualties changes the calculus for measuring military effectiveness. The size of the Persian opposition seems to vary considerably over the course of the march. Xenophon presents Mithridates' ranged troops growing from 400 (*Anab.* 3.3.6) to 4000 (3.4.2). Even if one of these numbers has been corrupted, the Persian ranged troops still considerably outnumber the 200 Rhodian slingers, and the losses for the Greek light-armed troops were significant: '50% of the light troops died on the campaign as opposed to 25% of the hoplites'.[50] Maintaining cohesion amidst such losses demonstrates the effectiveness of Xenophon's leadership and the specific motivating factors affording task cohesion. In contrast, Persians troops were much more risk-averse (*Anab.* 3.4.14) and did not need to engage if they would be endangered.

Conclusion

When the Greek army found itself in unfamiliar, irregular terrain, beset by skirmishers on all sides over a protracted period of time, a specialist unit of Rhodian slingers was formed. Though small in number, they regularly proved effective, as the Greek army marched toward the Black Sea. This discussion has aimed to rehabilitate the role of the ancient slinger through an examination of its most detailed surviving account.[51] It has proposed new interpretations of διαταχθέντες (3.4.15) and νεῦρα (3.4.17, see n. 15) as well as a deeper understanding that slingers brought a specific military expertise that was lacking among the initial deployment of the Ten Thousand. Xenophon's claims about the range of the sling are to be trusted, and they reflect the practical understandings of a skirmishing soldier, and not the idealized maximum ranges often isolated today. This understanding is reflected in *Anabasis* 3.4.15–17.

The introduction of a unit of Rhodian slingers improved the overall military performance of Xenophon's troops. It was defined in terms of an unstable shared ethnic identity, yet this provided a mechanism for creating unit cohesion, even when slingers themselves were dispersed through the ranks. Unit cohesion was established and maintained by tactical need, technological efficiency, and the realities of skirmishing. The *psiloi* among the Ten Thousand continued to employ three different types of missile weapon, as each technology when practiced by a specialist soldier served particular functions. Xenophon's integration of the humble sling recognized its value and his choice of slingers acknowledged both their expertise and their ability to work together within an improvised military unit.

Notes

1 For overviews of the sling in antiquity, see Echols (1950), Pritchett (1991): 1–67, Vutiropulos (1991), and, for a non-academic audience, Korfmann (1973) and Elliott (2008). Rihll (2009) and Kelly (2012) have helped revitalize the scholarly understanding of the sling. Buchholz (1965) traces the iconography of slinger. Except where noted, translations from Xenophon's *Anabasis* are taken from Brownson and Dillery (1998), which may be taken as neutral on the issues I wish to discuss; translations from Vegetius are from Milner (1996); others are my own. I would like to thank the editors of this volume for all their help and patience.

2 Anderson (1970): 115, 117; Huitink and Rood (2019): 140. Kelly (2012): 281 n. 13 notes the possibility that lead bullets were made and used by Persians too (see also n.47 below), although this is not clear from Xenophon's account. Sling bullets of any material consistently weigh only 30–40 grams; see Rihll (2009): 147 and 150 for the weights of Roman *glandes*.

3 Lee (2007): 55 n. 72 assumes the Rhodian recruits came from among the hoplites (indeed, for Rawlings (2000): 240 this is a mark of hoplite adaptability), rather than Whitby's (2004): 217–18, less plausible suggestion that they came from among the camp followers. While I agree with Lee, my conclusions do not depend on that assumption.

4 Mesopotamia, finds at Umm Dabaghiyah: Kirkbride (1974): 88; cf. Kelly (2012): 278–9. Syria, Tell Hamoukar: Lawler (2006): 1458 (unfired bullets deform on impact to create a bigger surface area, Reichel (2006): 72–5).

5 However, an apparent decline of the sling in early Iron Age Thessaly may be a result of the dependence on grave goods for interpreting that period; when artistic sources begin, there are clearer indications of slings and bows being used, Georganas (2005): 71.

6 Robinson (1941): 418–43, see also Lee (2001): 13 and Snodgrass (1967): 117.

7 See Völling (1990) and Rihll (2009). Trajan's column shows slingers holding ammunition in the fold of their cloak of their shield arm, Korfman (1973): 36; Kelly (2012): 197. For the staff-sling, see Livy 42.65.9–10; Polybius 27.11.1–7; Vegetius 1.16 and 3.14. Hollenback (2005) and (2009) discuss the dart sling.

8 Harrison (2006).

9 Similarly, Xenophon gives orders τοὺς γυμνῆτας λίθων ἔχειν μεστὰς τὰς διφθέρας (5.2.12, 'to the slingers to have their bags full of stones'). 'Slingers' here are τοὺς γυμνῆτας ('the naked ones'), referring to their lack of armour. Snodgrass (1967): 107 discusses fifth-century weapon costs.

10 For archers in Greek warfare generally, see Trundle (2010): 146–52; Huitink and Rood (2019): 135–6. Arrian relates how Alexander, on encountering Persian missile troops, recalled Xenophon's narrative (Arrian, *Anab.* 2.7.8). Alexander also used archers and slingers to provide cover for an advance (Arrian, *Anab.* 4.4.5, cf. Xen. *Anab.* 5.2.13–15), and see Anderson (1970): 115–17.

11 They might also have been heavier, but it is possible the Persians were using arrows made of reeds, as at Herodotus 7.76.1; see also Kelly (2012): 275.

12 Text is as in Brownson and Dillery (1998), accepting Matthiae's γε for τε. The OCT, following Madvig, marks the final three words as corrupt, but with Matthiae's correction further alteration is not needed (though see Huitink and Rood (2019): 155). Waterfield (2009): 71 expands the sense ('... because the Rhodians could hurl their sling-shot further than the Persians, and the Cretan archers could shoot further than their Persian counterparts'), but, in so doing, denies the special property of the sling. Warner (1972): 164, suggests that the slingers could shoot '... further even than most of their archers'; this interpretation emerges from the fact that Cretans could access Persian arrows (cf. Lendle (1995): 178) and presumes an impossibly static combat situation (on which. see below p. 000). Neither of these interpretations is needed.

13 Xenophon *Anab.* 3.3.7 equally observes that the javelins thrown by Greek peltasts also lacked the range of Persian archers and slingers. Peltasts were typically armed with thrown javelins, 3.3.8, cf. 1.10.7; Trundle (2010): 152–7; Parke (1933): 48–57. For modern experimentation determining javelin range, see Harris (1963); Hutchings and Brüchert (1997); Murray *et al.* (2010), (2011), (2012); Winter (2012). The javelin was typically thrown with the assistance of a thong (see *Anab.* 5.2.12). Harris (1963) describes some practical implications of throwing the javelin with the thong, distinguishing military from athletic endeavours; with refinements by Winter (2012).

14 Unlike bowstrings or lyre strings, where maintaining constant tension is required and so the use of sinew or gut is to be preferred, there is no similar need with a sling. The word νεῦρα can refer to the natural fibers in gut (treated sheep or goat intestines, typically) or the sinew of a tendon (*Iliad* 15.463, 469, 16.316, Xen. *Anab.* 4.2.28), but also to plant fibres (Plato, *Statesman* 280c). This last interpretation is to be understood here (the word is also used of slings by Quintus of Smyrna 11.112). Vegetius (3.14) says slings can be made of flax or hair, preferring the latter; in any case, the use of gut would be very unusual. For the opposite view, see Huitink and Rood (2019): 157.

15 The use of πλέκειν may also suggest the braiding or plaiting of vegetable fibers (see previous note). In Homer, Locrian slingers use plaited wool slings, where dense volleys can cause formations to break (*Iliad* 13.712–18). Livy 38.29.6 describes the elaborate slings of the Balearic islanders. The offer of cash perhaps indicates that no one among the Ten Thousand was receiving a daily wage at that point, Huitink and Rood (2019): 141.

16 Bosman (1995); Baatz (1990). Kelly (2012): 282, describes the indications of last minute and informal casting, as Roman soldiers cast bullets from thumb-holes made in the soft ground. Lead bullets continued to be used to end of the second century CE, *Life of Severus* 11.3; Kelly (2012): 279–80. Significantly lead does not figure in the references to slings in Vegetius, by which point natural or ceramic bullets again become standard, Rihll (2009): 147. Even when the technology to

make lead bullets existed, soldiers might continue to use other materials because access to lead could not be guaranteed. Xenophon *Anab.* 3.4.17 stresses that lead was available to his troops in the local villages, even if it was being put to another use.

17 Rihll (2009): 160–4, argues implausibly that bullets are fired from hand-catapults (i.e. what in North America is called a slingshot), but this would mean that νεῦρα is to be understood as sinew. Xenophon is discussed explicitly at 164–65, where it is suggested the Rhodian slingers were using stone-bows (i.e. bullet-shooting crossbows).

18 Lee (2001): 16, 20–1.

19 Richardson (1998): 45; Lee (2001): 16.

20 Korfman (1973): 36.

21 It is not necessary to argue that island life was congenial to the development of effective slinging, though that assumption did exist in antiquity: Strabo 3.5.1.167–68 describes the sling's role in the upbringing of children on the Balearic Islands. For Balaearic slingers, see Florus 1.43.8; Livy 38.29.5; Vegetius 1.16; Ovid, *Met.* 4.709–10. See Diodorus of Sicily 5.18.3–4, and 15.85.5 for slings among Thessalian children. Livy mentions Cyrtian slingers (37.40.9–14) and Achaean slingers (38.29.8). Cretan slingers are discussed by Kelly (2012).

22 MacCoun and Hix (2010): 137–8.

23 See summaries at MacCoun and Hix (2010): 139–40; Armstrong (2016): 101–2.

24 MacCoun (1993): 291.

25 See MacCoun and Hix (2010): 155, with bibliography.

26 As Armstrong (2016): 104, notes, 'in many classical armies, membership in the state seems to have been the only overarching, unifying bond which all the soldiers initially shared'.

27 'All of the evidence indicates that military performance depends on whether service members are committed to the same professional goals', MacCoun, Kier, and Belkin (2006): 652. I would suggest that the mercenaries' task cohesion fragments in books 5 through 7 as they travel along the coast of the Black Sea, after the emotional high point of 'The Sea! The Sea!' (*Anab.* 4.7.24), which, in one way, represents the fulfillment of the task unifying the Ten Thousand.

28 MacCoun and Hix (2010): 156.

29 MacCoun and Hix identify this as 'group pride', which is not dependent on direct knowledge of other group members, (2010): 140, with bibliography.

30 Xenophon's details concern the rearguard primarily, where he was in charge and possessed eyewitness knowledge. For the command of the rearguard at this point of the campaign, see Roy (1968).

31 The deletion of Σκύθαι by Krüger has been generally accepted; these are the Cretan bowmen already mentioned. Because Scythians fought on horseback, they employed short bows, as the Cretans do here, McLeod (1965): 2, 14–15, and (1968); the intrusive word possibly derives from a gloss that reflects this knowledge. Such an interpretation is preferable to the assumption that Xenophon was somehow careless or forgetful, a view still current that this chapter rejects: e.g. Rihll (2009): 165, 'Close examination of the passage suggests that Xenophon's memory was imperfect (e.g. the Cretan archers of 3.3.15 are Scythian archers by 3.4.15). This should not surprise, for he may have been writing this account some 30 years after the events'.

32 Huitink and Rood (2019): 154: 'the δια- prefix implies distribution at intervals to increase the spread of missiles (by contrast to the enforced shooting form within the square at 3.3.7)'. This is also suggested by Brownson's 1922 translation of *Anab.* 3.3.18 τῷ σφενδονᾶν ἐν τῷ τεταγμένῳ ἐθέλοντι, 'for the man who will volunteer to serve as a slinger at his appointed post' (but not Brownson and Dillery (1998): 261: 'for the one assigned to be a slinger').

33 MacCoun, Kier, and Belkin (2006): 649.
34 MacCoun and Hix (2010): 141.
35 Fraser and Bean (1954).
36 Hanson (2009): 15–16; Trundle (2010): 144–5.
37 Grossman provides a useful starting point, discussing combat psychology generally for a popular audience. He describes the choices a soldier faces when threatened (fight or flight, posture or submit; see (2009): 5–17), but notes that, traditionally, combat was almost never to the death (6). As Krentz (1985): 18–19, demonstrates, typical casualty rates in Classical warfare were roughly 5% for the victors and 14% for the losers. Even if light-armed troops are more vulnerable (or more expendable) than hoplites, the foundation for Grossman's assumptions appear to hold for antiquity. Grossman (2009): 30–7, further identifies the resistance of the attacker to the act of killing; but distance is important in reducing this resistance (97–106); such distance can be physical and emotional (156–70).
38 Hanson (2009): 117–25.
39 Echeverría (2012): 304–8, notes other ways in which Xenophon's description of phalanx deployment is irregular.
40 Plato, *Laws* 625d, cf. Thuc. 4.32–37; Xen. *Anab.* 4.2.28; Kelly (2012): 276.
41 For the javelin: Harris (1963); and Hutchings and Brüchert (1997): 892, give ranges c. 20–25 m (though Winter (2012): 203, claims up to 33 m as 'typical'). For archery: McLeod (1965): 8, suggests 50–60 m for an accurate shot, though offering an extreme limit of 'perhaps about 175 metres' (15). Drawing on Vegetius 2.23, who says *fundatores* set targets 600 feet away, many believe the sling also habitually reached 175–180 m, though Korfman (1973): 37, provides anecdotal evidence for range of up to 240 m from inexpert slingers. Huitunk and Rood (2019): 140, suggest a much greater range, up to 365 m for slings and 180 m for most Greek bows, with the Cretans able to achieve 230 m (136). Such extreme ranges, even if technically possible, would not yield any accuracy, and I suspect the effective ranges are less than half of these estimates.
42 Lead missiles have been found with the message ὗσε ('it's raining [sc. with bullets]'; Kelly (2012): 299–300.
43 Harrison (2006).
44 Slingers could also prove effective when used against cavalry (Thuc. 6.22.1). Hanson (2009): 140 notes that, 'Running the last two hundred yards of no-man's-land, as the hoplites at Marathon showed, limited such exposure to attack [from missiles] until the protective cover of the general mêlée could be reached' (cf. Arrian, *Anab.* 2.10.3).
45 A preliminary announcement that some Roman sling bullets may have made a sound as they flew indicates another dimension to the terror of facing a unit of slingers, Metcalfe (2016). YouTube videos of modern whistling sling bullets provide some indication of the potential psychological impact, which combined with the difficulty of seeing bullets in flight could prove very effective, Canal de hondero191 (2012a and 2012b).
46 Huitink and Rood (2019): 140. In time, messages were introduced as part of the process of manufacturing lead bullets, since short messages or designs could be cast in relief (see above, n. 43). Kelly (2012): 295–8, argues that the principal benefit of such messages was to encourage the attackers, reinforcing unit cohesion through the presence of brief taunting messages, rather than to demoralize the defender, cf. Lee (2001): 276. For Kelly (2012): 281–82, the text is a key component of the casting (see 282–96 for a variety of inscriptions, with taunts at 290–6).
47 Kelly (2012): 281; cf. the fight against Persians at 3.3.7–11.

48 This point is made by Lindybeige (2010a), a non-academic ancient weapons enthusiast who draws on a range of practical experience. See also Lindybeige (2010b) for technique when using a sling. These are rich considerations of the practical implications of the sling in battle.

49 See Hanson (2009): 210–18, for wounds in battle generally. Vegetius 1.16 insists that 'smooth stones shot with a sling or sling-staff are more dangerous than any arrows, since while leaving the limbs intact they inflict a wound that is still lethal, and the enemy dies from the blow of the stone without loss of blood'. Celsus 7.5.5 also introduces the possibility of poisoned missiles, but this is not likely to refer to sling bullets.

50 Trundle (2010): 160, citing Best (1969): 78.

51 Trundle (2010): 144: 'Slingers, though skilled and effective, received the least respect of the various specialist light troops available in the Greek world'.

Bibliography

Anderson, J. K. 1970: *Military Theory and Practice in the Age of Xenophon*, Berkeley.

Armstrong, J. 2016: 'The Ties that Bind: Military Cohesion in Archaic Rome'. In J. Armstrong (ed.), Circum Mare*: Themes in Ancient Warfare* (*Mnemosyne,* Supplement 388), Leiden: 101–119.

Baatz, D. 1990: 'Schleudergeschosse aus Blei: Eine waffentechnische Untersuchung', *Saalburg Jb* 45: 59–67.

Best, J. G. P. 1969: *Thracian Peltasts and their influence on Greek warfare*, Groningen.

Bosman, A. V. A. J. 1995: 'Pouring Lead in the Pouring Rain: Making Lead Slingshot under Battle Conditions', *Journal of Roman Military Equipment Studies* 6: 99–103.

Brownson, C. L. and J. Dillery, 1998: *Xenophon: Anabasis* (Loeb Classical Library), Cambridge, MA.

Buchholz, H. G. 1965: 'Die Schleuder als Waffe im Aegaeischen Kulturkreis', *Anadolu Arastirmalari* 2: 133–159.

Canal de hondero191. 2012a: 'Sling whistling projectiles', *Youtube.com*. Retrieved from: https://www.youtube.com/watch?v=svdL-kszLaw.

Canal de hondero191. 2012b: 'Sling whistling projectiles (part 2)', *Youtube.com*. Retrieved from: https://www.youtube.com/watch?v=svdL-kszLaw.

Echeverría, F. 2012: 'Hoplite and Phalanx in Archaic and Classical Greece: A Reassessment', *CP* 107: 291–318.

Echols, E. C. 1950: 'The Ancient Slinger', *The Classical Weekly* 43: 227–230.

Elliott, P. 2008: 'Humble and Deadly: The Ancient Slinger', *Ancient Warfare* 2 (1) (February/March): 24–27.

Fraser, P. M. and G. E. Bean, 1954: *The Rhodian Peraea and Islands*, London.

Georganas, I. 2005: 'Weapons and Warfare in Early Iron Age Thessaly', *Mediterranean Archaeology and Archaeometry* 5 (2): 63–74.

Grossman, D. 2009: *On Killing: The Psychological Cost of Learning to Kill in War and Society*, Rev. ed., New York.

Hanson, V. D. 2009: *The Western Way of War: Infantry Battle in Classical Greece*, 2nd ed., Berkeley.

Harris, H. A. 1963: 'Greek Javelin Throwing', *G&R* 10: 26–36.

Harrison, C. 2006: 'The sling in Medieval Europe,' *Bulletin of Primitive Technology* 31. Retrieved from: http://www.chrisharrison.net/index.php/Research/Sling.

Hollenback, G. M. 2005: 'A New Reconstruction of the *kestros* or *cestrosphendone*', *Arms & Armour* 2: 79–86.

Hollenback, G. M. 2009: 'Polybius' Description of the *kestros*', *Mnemosyne* 62: 459–463.

Huitink, L. and T. Rood, 2019: *Xenophon:* Anabasis *Book III*, Cambridge.

Hutchings, W. K. and L. W. Brüchert, 1997: 'Spearthrower Performance: Ethnographic and Experiential Research', *Antiquity* 71: 890–897.

Kelly, A. 2012: 'The Cretan Slinger at War – A Weighty Exchange', *ABSA* 107: 273–311.

Kirkbride, D. 1974: 'Umm Dabaghiyah: A Trading Outpost?', *Iraq* 36: 85–92.

Korfmann, M. 1973: 'The Sling as a Weapon', *Scientific American* 229 (October): 34–42.

Krentz, P. 1985: 'Casualties in Hoplite Battles', *GRBS* 26: 13–20.

Lawler, A. 2006: 'North Versus South, Mesopotamian Style', *Science* 312: 1458–1463.

Lee, J. W. I. 2001: 'Urban Combat at Olynthos, 348 BC'. In P. W. M. Freeman and A. Pollard (eds.), *Fields of Conflict: Progress and Prospect in Battlefield Archaeology* (BAR International Series 958), Oxford: 11–22.

Lee, J. W. I. 2007: *A Greek Army on the March: Soldiers and Survival in Xenophon's* Anabasis, Cambridge.

Lendle, O. 1995: *Kommentar zu Xenophons Anabasis (Bücher 1–7)*, Darmstadt.

Lindybeige (Nikolas Lloyd). 2010a: 'Some more points about slings', *Youtube.com.* Retrieved from: https://www.youtube.com/watch?v=sGSsbCPeocU.

Lindybeige (Nikolas Lloyd). 2010b; 'Slinging techniques', *Youtube.com.* Retrieved from: https://www.youtube.com/watch?v=yJ3bBkRIJNU.

MacCoun, R. J. 1993: 'What Is Known about Unit Cohesion and Military Performance'. In *Sexual Orientation and U.S. Military Personnel Policy: Options and Assessment*, Santa Monica: 283–331.

MacCoun, R. J. and W. M. Hix, 2010: 'Cohesion and Performance'. In B. Rosteker (ed.), *Sexual Orientation and U.S. Military Policy: An update of RAND's 1993 study*, Santa Monica: 137–158.

MacCoun, R. J., Kier, E. and A. Belkin, 2006: 'Does Social Cohesion Determine Motivation in Combat? An Old Question with an Old Answer', *Armed Forces & Society* 32: 646–654.

McLeod, W. 1965: 'The Range of the Ancient Bow', *Phoenix* 19: 1–14.

McLeod, W. 1968: 'The Ancient Cretan Bow', *Journal of the Society of Archer-Antiquaries* 11: 30–33.

Metcalfe, T. 2016: 'Whistling Sling Bullets Were Roman Troops' Secret "Terror Weapon"', *Livescience.com*. Retrieved from: http://www.livescience.com/55050-whistling-sling-bullets-from-roman-battle-found.html.

Milner, N. 1996: *Vegetius: Epitome of Military Science*, 2nd ed., Liverpool.

Murray, S. R., Sands, W. A., Keck, N. A., and D. A. O'Roark, 2010: 'Efficacy of the *Ankyle* in Increasing the Distance of the Ancient Greek Javelin Throw', *Nikephoros* 23: 43–55, 329–33.

Murray, S. R., Sands, W. A., and D. A. O'Roark, 2011: 'Throwing the Ancient Greek *Dory*: How Effective Is the Attached *ankyle* at Increasing the Distance of the Throw?', *Palamedes* 6: 137–151.

Murray, S. R., Sands, W. A., and D. A. O'Roark, 2012: 'Recreating the Ancient Greek Javelin Throw: How Far Was the Javelin Thrown?', *Nikephoros* 25: 143–154.

Parke, H. W. (1933). Greek Mercenary Soldiers, from the earliest times to the battle of Ipsus, Oxford.

Pritchett, W. K. 1991: *The Greek State at War, Part V*, Berkeley.
Rawlings, L. 2000: 'Alternative Agonies: Hoplite Martial and Combat Experiences beyond the Phalanx'. In H. van Wees (ed.), *War and Violence in Ancient Greece*, London: 233–259.
Reichel, C. D. 2006: 'Hamoukar', *Oriental Institute Annual Report 2005–2006*, Chicago: 65–77.
Richardson, T. 1998: 'The Ballistics of the Sling', *Royal Armouries Yearbook* 3: 44–49.
Rihll, T. 2009: 'Lead "slingshot" (*glandes*)', *JRA* 22: 147–169.
Robinson, D. M. 1941: *Excavations at Olynthus, Part X: Metal and Minor Miscellaneous Finds*, Baltimore.
Roy, J. 1968: 'Xenophon's *Anabasis*: The Command of the Rearguard in Books 3 and 4', *Phoenix* 22: 158–159.
Snodgrass, A. M. 1967: *Arms and Armour of the Greeks*, London.
Trundle, M. 2010: 'Light Troops in Classical Athens'. In D. M. Pritchard (ed.), *War, Democracy, and Culture in Classical Athens*, Cambridge: 139–160.
Völling, T. 1990: 'Funditores im römischen Heer', *Saalburg Jb* 45: 24–58.
Vutiropulos, N. 1991: 'The Sling in the Aegean Bronze Age', *Antiquity* 65: 279–286.
Warner, R. 1972: *Xenophon: The Persian Expedition* (with introduction and notes by G. Cawkwell), Harmondsworth.
Waterfield, R. 2009: *Xenophon: The Expedition of Cyrus*, Oxford.
Whitby, M. 2004: 'Xenophon's Ten Thousand as a Fighting Force'. In R. Lane Fox (ed.), *The Long March: Xenophon and the Ten Thousand*, New Haven: 215–242.
Winter, T. 2012: 'How to Throw a Spear on a Sling', *Ancient World* 43: 203–211.

3 Keeping It Together: Aeneas Tacticus and Unit Cohesion in Ancient Greek Siege Warfare

Aimee Schofield

According to Aeneas Tacticus, the author of our earliest surviving military treatise, the key to surviving a siege in the world of fourth century BC Greece was *homonoia*, or unanimity. This sense that maintaining cohesion (i.e. a shared sense of purpose in defending the city) in both the social order of the city (social cohesion) as well as among the fighting forces (task or mission cohesion) is repeated throughout his *How to Survive Siege Warfare*. The concept of *homonoia* is crucial to understanding not only how the Greeks thought about siege warfare, but also how they defended their cities and homes when they came under attack.

However, the demands placed on the fighter came from more than just his native city. Sieges create a highly unusual and complex set of circumstances in warfare, where the lines between the 'civilian' and military spheres become blurred. Those normally considered to be non-combatants[1] are closely involved both with the fighting and the fighters in ways we would not expect in the set-piece battles fought by armies. Soldiers remain in close proximity to their families – their wives, mothers, sisters, fathers, and children; these family members may also take part (directly or indirectly) in the fighting itself. As a result, a male citizen of fighting age would be presented with competing priorities: was his first duty to his family or his city?

This chapter builds on the concepts developed in Crowley's study of the psychology of the Athenian hoplite, which explains how unit cohesion was maintained among Athenian forces.[2] Where this chapter differs significantly, however, is that it will attempt a much broader view of unit cohesion which covers the wider Greek world. In *How to Survive Siege Warfare*, Aeneas Tacticus deliberately distances himself from writing about a specific city, focusing instead on a purely hypothetical city-state: his aim is to write a treatise which is applicable to any city, regardless of political orientation.[3]

Moreover, whereas the fighting Crowley describes takes place away from the city boundaries, this chapter will highlight how competing pressures on the fighters were significantly stronger in sieges, and the ways in which Aeneas Tacticus believes that these pressures can be combatted. This also allows this chapter to consider the effects which women, children, slaves,

DOI: 10.4324/9781315171753-4

and those too old or infirm to fight had on the cohesion of the military under siege conditions.

Due to the nature and range range of advice contained within *How to Survive Siege Warfare*, evidence for the military unit is very limited (to the point where it is impossible to tell the size and makeup of the military unit beyond the immediate primary group). Nevertheless, three major aspects of Crowley's model will be considered: the role of the primary group in unit cohesion,[4] for which we have the most evidence from Aeneas Tacticus; the role of the socio-political system,[5] for which we have less evidence (since Aeneas is working from a hypothetical viewpoint); and, thirdly, the compliance relationship,[6] for which we only have scattered evidence in *How to Survive Siege Warfare*.

The Ancient Greek Siege

The ancient Greek 'city' ranged in size massively, from what today would be considered hamlets to those with populations of many thousands. Different political systems and allegiances dominated within individual cities over time. As a result, there was no such thing as a typical city; nor could there be a typical siege. However, there were some significant commonalities in the ways in which a besieger approached his task.

By far the most common approach to attacking a city was an attempt to cut it off from the outside world. Blockades[7] and circumvallation[8] were popular methods; direct attacks on the walls of the city were rare until the catapult became a standard part of the military arsenal.[9] Another standard tactic in this period was the targeting of property outside the city, and ideally within view of its walls.[10] Devastation of the countryside was partly a method of foraging, partly a way to keep troops occupied, and partly a way to taunt those property owners who had been evacuated to the city proper. In effect, the attacking force was daring those in the safety of the city to come out and defend their homes and their livelihoods.

It might be expected that those within a city would automatically band together to defend their homes. After all, it could be argued that a common threat and the risk of death or enslavement were natural motivators that would bring people together. Our texts from this period show that this was not the case. Aeneas Tacticus points to potential fault lines between the rich and poor, which had to be settled in the event of a siege to avoid internal unrest.[11] Moreover, political factionalism was rife in ancient Greek cities even in times of peace; were an oligarchic state to attack a democratic one (or vice versa) proponents of the attackers' political system could be expected to support them from within the city walls.[12] The larger cities also had substantial *metic* (non-citizen/foreigner) populations which could be considered at risk of forming a fifth column in times of war.[13] And then, if an attacking force were to offer safe passage or other incentives to a traitor (and potentially his family too), there was risk to the city from the individual.

The difficulties of life under siege could also cause pressures within the city to mount. Owing to the nature of siege warfare in the fifth and early fourth centuries (with attacking forces making significant efforts to cut off the city's supply lines) there was a severe risk of shortages.[14] Black markets and inflation of prices could cause further tensions. The privations of life under siege no doubt also contributed to everyday discomforts and potentially led to dissatisfaction with the city's leadership. Here in particular, it is worth considering the position of the women and children in the besieged city and their influence on those actively defending it. Whether the women were able to maintain their civic unity under these conditions or whether panic, distrust, and fear took over could affect the outcome of the siege – as the attacking forces intended. The messages that the fighter received at home could directly conflict with those he received from his superiors, affecting his morale and commitment to the city's leaders. The extent to which this played out in real life is unknown; as with much of the domestic sphere in ancient Greece, this largely goes unrecorded.[15]

It is certainly the case, then, that cohesion within the city could not be taken for granted. Civil uprisings, political dissent, fifth columns, and individual treachery all posed a very real risk to the city under siege. When looking at history through the lens of hindsight, scholars frequently expect their subjects to have behaved logically and with insight. However, in a siege situation, the competing pressures placed upon the individual of family, political affiliation, and city were such that this cannot safely be assumed. It is in this climate that Aeneas Tacticus' *How to Survive Under Siege* emerges to advise cities on precautions against such internal threats.

How to Survive Siege Warfare and the City

How to Survive Siege Warfare gives us an image of siege warfare in the second quarter of the fourth century BC. Dating the text is difficult, and, based on the tactical exempla given by Aeneas throughout the work, scholars have settled on dates of composition with a *terminus post quem* of 360 BC[16] and a *terminus ante quem* of 346 BC.[17] Whitehead has convincingly argued that the date of composition lies in the late 350s or: 'perhaps very close to 355 itself'.[18] At this point, catapults and other siege machinery (other than rams) are seen as a novelty,[19] which indicates a publication date early within this range, since there is strong evidence for catapults being widespread in the Greek world by the mid-fourth century.[20]

Aeneas is explicit in the fact that his intended audience is very broad. He is not writing for a particular type of city (in terms of size, political structure, or geographic location); instead, his work is intended to be as generic as possible so that it could be applied by any of his readers to any city they might have to defend.[21] Aeneas refers to what states who have particular assets or terrains (or, indeed, if they lack them) may be able to accomplish as examples, but never suggests that a city must have these assets in order to implement his more general advice.[22] Aeneas' primary concern is the maintenance of the *status quo*, whatever that might be in any given city.[23] Examples of previous tactics used by both attacking and defending forces are given throughout *How to Survive Siege Warfare* and are taken from cities and other sites scattered broadly around the Greek world, without focusing on any particular type of city.[24]

The advice on unit cohesion in *How to Survive Siege Warfare* is focused on Aeneas' major guiding principle. First and foremost in his mind, is the need for unanimity both broadly in the wider city (including the non-combatants within the population) and among the smaller units deployed in its defence. Given that sieges provide an unusual overlap between the civilian and military spheres, it is unsurprising that the approaches he recommends for fostering *homonoia* cover tactical, political, and social aspects of military activity and city life. These approaches to the maintenance of stability within the city and its military units can be split into three major themes: morale, since troops who are disheartened or dissatisfied are more likely to fail the city (either through treachery or lack of attention); status, since Aeneas believes that elite members of the city (i.e., those: 'with a surety in the *polis*') make the best leaders and fighters;[25] and proximity, since it is more efficient to group troops according to where they live in the city and their closest defensive landmarks (e.g., the city's *agora*). These themes each affect the cohesion of the military unit in different ways and at different levels. Taking Crowley's model as a basis for analysing the advice given in *How to Survive Siege Warfare*, in particular the effects of the primary group, the socio-political system, and the compliance relationship, this chapter will examine how unit cohesion could be created and maintained in the volatile atmosphere of a besieged fourth-century Greek city.

We're All in This Together? The Primary Group and Military Unit in Siege Warfare

One of the main concepts applied to Athenian hoplites in Crowley's work on the psychology of hoplite warfare is that of the primary group, i.e., the specific group in which an individual has intimate interactions with, such as: 'his or her family, work colleagues or friends'.[26] In military units, he points out, the primary group is much more identifiable than in a civilian context, and that this primary group can be found in the smallest subsections of a military unit, for example: 'a soldier's rifle section or squad'.[27] The soldiers within these primary groups support each other in numerous ways and work together to survive the engagements in which they are involved.[28]

The problems associated with establishing just what one's primary group is under siege conditions – as opposed to during set piece battles and warfare at a distance from the city – are highlighted by Aeneas Tacticus from the beginning of *How to Survive Siege Warfare.* In the preface to the work, he points out that those defending the city are acting on behalf of their gods, their homeland, their parents, their children, and other unspecified things.[29] The men fighting faced the total destruction of everything to which they had any loyalty,[30] and yet these same things were in competition for their attention during the siege. Aeneas recommends that those in positions of responsibility should have something to risk should the city fall, specifically their wives and children.[31]

At the same time, fighters were to be stationed as close to their homes as possible[32] not only to ensure that they would reach their assigned positions quickly, but also to ensure that they could: 'control those at home, that is their children and their wives'.[33] The fact that Aeneas recommends that men be stationed so close to home to keep an eye on their family members implies either that without such proximity the women and children would get in the way of the military, or that men might be so concerned about their family's wellbeing that they would risk neglecting their duties. In a siege situation both are plausible and Aeneas does not specify which of these he considers to be of greatest concern, or whether both possibilities are equally problematic during a siege. In any case, the presence of these non-combatants evidently affected the group dynamics of the military primary group.

Close links to family members could disrupt this primary group's effectiveness and solidarity even if they were not physically present in the city. A notable risk for dissent highlighted by Aeneas arises where hostages have been given and are held by the enemy.[34] The relatives of the hostages should be removed from the *polis*, according to *How to Survive Siege Warfare*,[35]or at the very least be kept occupied and away from anything too sensitive, to prevent them from causing trouble or betraying the city in the hope of generating some protection for their family member.[36] Family members in exile, too, risked exacerbating pre-existing factional tensions and undermining the cohesiveness of the defensive force.[37]

Unit cohesion could not, therefore, depend significantly on loyalty to the primary group in the way we might expect from modern military personnel and ancient soldiers fighting at a distance from their city. The relationships at play within the besieged city were considerably more complex and layered, and the presence of women, children, and other traditional non-combatants close to, or at the centre of, the fighting put the cohesion of the primary group at considerable risk of collapse. Aeneas does suggest ways to mitigate this challenge to unit cohesion within the primary group. In the example discussed above, where defensive forces are stationed close to their homes,[38] there would be some significant factors to support unit cohesion. As well as getting soldiers to their posts quickly and ensuring domestic order, assigning men to primary groups according to the neighbourhood in which they lived meant that units would be made up of men who most likely knew each other well, interacted with each other on a daily basis, and had a shared sense of identity.[39]

In Athens, recruitment and deployment were focused around a political system based on the *demes* and *trittyes* which provided the men serving in Athenian hoplite units with a shared sense of affiliation.[40] As Aeneas is writing for a general audience, he cannot assume that his readers live in cities with identical or similar mobilisation patterns, and he cannot assume anything about the political organisation of the city-state.[41] However, he can circumvent this problem by making use of the neighbourhoods of the hypothetical city to the same effect as achieved by the Athenians through their own political system.

Aeneas also advised on economic methods to foster unanimity in the besieged city. Sadly, the in-depth explanation of how to accomplish these goals were contained in another work of his, *Procurement*, which is no longer extant.[42] The basic premise of the measures was that of debt relief and handouts of basic necessities to the poorest, on the grounds that this would ensure that the poor did not feel overly burdened by both their existing financial problems and the strains brought about by the siege.[43] This would (in theory) help to alleviate the problems caused by supply issues, as well as reducing civil tensions caused by one social class being able to cope better with the siege than another. As with stationing men near their homes, this measure acknowledges the fact that soldiers' primary groups are much broader than just their military units and highlights concerns they must have had not only for the physical safety of their families, but also for their need to ensure that the family as a whole could survive financially. Such a measure could help to bring the city as a whole together, as well as the individual fighting unit.

This raises an interesting question about the social standing of those fighting to defend the *polis*. In conflicts outside the city walls, wealth was a determining factor over who could fight, with property qualifications determining a citizen's role as a cavalryman, infantryman, or sailor.[44] On the other hand, by the mid-fourth century BC, groups of the more lightly (and inexpensively) armed peltasts were much more common on the battlefield.[45] In addition, in a siege situation commanders could not afford to be too particular in their choice of soldiers to defend the city, given the likely limited resources at their disposal. Aeneas even suggests that cities might need to make use of women to make the numbers of defenders seem greater.[46]

Having men from different socio-economic levels fighting together while defending the city, when in all likelihood they would normally fall into different primary groups when fighting away from the city, must surely have had implications for the cohesiveness of their fighting units. It is clear from the advice in *How to Survive Siege Warfare* that Aeneas expects the richest or most elite men in the city to take on leadership roles,[47] alongside those with significant military experience,[48] who were likely themselves part of the elite of the city. Those in charge of the groups allocated to the areas close to their own homes, however, were simply to be the: 'most capable and most sensible in each street',[49] with the reader left to make up their own mind over how these men should be chosen. In any case, a system set up to group soldiers according to their location, rather than the units in which they might normally have expected to find themselves, may well have affected the cohesion of the military primary group in these circumstances.

The one set of people whose primary group was purely military presented significant problems to the city's leadership, according to Aeneas Tacticus. Mercenaries were in all probability a group of men from a number of different cities, who as a group owed no loyalty to the city which hired them beyond the fee paid.[50] As a result, Aeneas – mistrusting almost everyone even at the best of times[51] – advises his readers on how to ensure that, firstly, the mercenary forces hired by the city are paid[52] and do their job in defending the city well and, secondly, that they have no opportunity afforded to them which would allow them to betray the city to the enemy or switch sides.

A policy of divide and rule was to be followed, according to Aeneas' advice, with mercenaries housed separately in small groups.[53] They were to be kept separated from each other at all times, except while on guard duty or performing other necessary tasks. This would limit their exposure to their normal primary group, making it harder for them to foment trouble. Unfortunately, limited interaction with their primary group might serve to weaken their personal ties and thus inhibit their ability to work together in combat if they were separated over a long period during the siege. By contrast, an earlier passage in *How to Survive Siege Warfare*, in which Aeneas advises on ways to discipline mercenaries should they later defect to the enemy, implies that mercenaries were commonly encamped together and not separated,[54] thus allowing them to maintain close ties with their primary groups.[55] Like the other advice contained within *How to Survive Siege Warfare*, that on how to accommodate mercenary troops is evidently varied and adaptable to the differing circumstances in a range of cities.

In short, the primary group was of key importance to the cohesion of a variety of units within the besieged city. However, Aeneas saw that a range of factors could affect the efficiency and cohesion of the military primary group. The presence of family members, especially women and children, in the besieged city, or their use as hostages to an enemy state, or indeed family members living in exile who disapproved of the city's *status quo* could put the security and smooth running of military operations at significant risk. Primary groups within the city were not necessarily formed in the same way as those for operations undertaken outside of the city. Mercenaries, where possible, were to be denied opportunities to maintain links with their primary groups in order to prevent them from initiating subversive activity. While Aeneas offers some advice to mitigate these problems, it is clear that the primary group simply could not be relied upon as a basis for unit cohesion during sieges in the way it could in more traditional forms of warfare.

What Are We Fighting For? The Socio-Political System

As has been outlined above, Aeneas Tacticus did not write for any one political system. This subject has been discussed frequently in scholarly writing, and this chapter has already reiterated many of the key points in this area.[56] However, although Aeneas singles out no specific political standpoint as his target audience, the ideology and political status of the city-state are key to his advice on creating cohesion among the fighters defending it.

In Crowley's model, it is the Athenian socio-political system which forms the basis of the analysis;[57] the hypothetical approach which Aeneas takes is considerably broader and, as a result, the model must therefore be adjusted to accommodate this. Athens was considerably larger (both geographically and in terms of population) than some of the cities to which Aeneas was targeting his advice.[58] The hypothetical city could be governed by an oligarchy or a tyranny, though, like the Athens of Crowley's analysis, the city could also be democratic. Whereas citizens in Athens shared a common bond, through the ability to take an equal part in the democratic institutions of the city,[59] this cannot be assumed for Aeneas' target audience.

Maintenance of the political *status quo* in the city is crucial to its survival in a time of siege, according to Aeneas.[60] His biggest cause for concern – that the city might fall to treachery, or that the city might enter a state of civil disturbance (*stasis*) – could be mitigated through measures taken to ensure that the existing regime remained in power during the siege. Ensuring that the people within the city remained content enough with the *status quo* that they would see no need to challenge it or betray the city to an external force was also essential. As a result, the bulk of *How to Survive Siege Warfare* deals in one way or another with ensuring that no fifth columns have the opportunity to develop within the hypothetical city.

Some of the measures which Aeneas recommends appear benign at first glance. The idea of debt relief, discussed above, would certainly have had benefits for the poorest (and probably largest) group of people within the city.[61] However, Aeneas does not tell the readers of *How to Survive Siege Warfare*[62] how this could be achieved without alienating or antagonising the richest members of the city, who were likely more influential and better resourced to stage a coup if they were so inclined. This is left to his book *Procurement* which, as noted above, is no longer extant. Militarily, this measure would bring together the lowest positioned soldiers, and the benefit which debt relief conferred on them would likely enhance their loyalty to the city; it could, though, if handled poorly, serve to drive a wedge between the people and their commanders (i.e., the poor and the rich), thus creating problems in the cohesion of the defensive forces as a whole.

In Athens, distrust or envy between the poor and the rich could be offset politically (to some degree) by the system of liturgies, the ability for both rich and poor to serve as jurors, and the fact that many major political offices were temporary and chosen by lot.[63] Festivals and processions, political institutions, and the *ephebeia* served to bring Athenian citizens together socially, politically, and militarily.[64] These promoted norms to the Athenian citizen body and helped it to remain cohesive for much of the fifth and fourth centuries BC.[65]

However, the advice in *How to Survive Siege Warfare* highlights the possibility – or even the likelihood – that many cities were not as united as Athens. Moreover, some of the institutions and activities which helped Athenian citizens to form close bonds were to be regarded as potential opportunities for coups or betrayal. Festivals in particular were risky for the city since the authorities (and the main citizen body) were likely to be off their guard and one faction might have the opportunity to overthrow another.[66] City leaders were advised to send potential conspirators home to celebrate festivals, rather than allowing them to take part with the rest of the community; Aeneas suggests that this would not only remove the danger of their committing treacherous acts during the celebrations, but would make these men feel that they had been singled out for a particular honour (and therefore not feel slighted by being sent away from their posts).[67] Factionalism is certainly evidenced at such celebrations, demonstrating that unanimity in the city could not be taken for granted.[68]

Some of Aeneas' proposals for the maintenance of the *status quo* appear to be significantly less benign. If the city were in danger of being besieged, Aeneas recommends getting rid of any political opponents of the current regime. In particular, he suggests that it is wise to remove those who have been leaders in the city before. These men should be sent as embassies away from the city to prevent them from having the chance to cause trouble.[69] In a more troubling piece of advice, Aeneas proposes that money should be offered to informants for denouncing conspirators.[70] He does not go so far as to suggest how the veracity of the information provided this way should be confirmed, nor does he advise his readers on what to do with any suspected traitors discovered in this way. It is possible that such moves might serve to bolster morale (and thus further promote unit cohesion) by reassuring the soldiery and the population that any risk of treachery was being dealt with. However, evidence from more recent history suggests that such denunciations are more likely to lead to further suspicion and a lower morale in the groups affected, potentially making the bonds of the military units weaker, while at the same time exacerbating existing conflicts within the population.[71]

A further problem, in Aeneas' mind, was that the socio-political backgrounds of the defenders could often be radically different. In the context of siege warfare, a defensive force might include troops from several cities, or even mercenary units made up of men from around the Greek world. Cultural identifiers (including having different names for the same gods and differences in the dialect spoken) could lead not only to cultural misunderstandings, but more troublingly to the miscommunication of passwords.[72] Aeneas advises, therefore, that the choice of passwords should be: 'very common to everyone',[73] in order to avoid potential confusion. In addition, it was entirely possible that the groups making up these mixed forces would have very different political standpoints and military objectives.[74] Mercenary troops by their very nature had no fixed loyalty, and Aeneas even advises that they should be given an opportunity to leave before a siege began, and that anyone who caused trouble after that point should be subject to the death penalty.[75] He also advised that the city should never allow an allied force stronger than the city's own army to enter the city limits in order to reduce the risk that those the city believed were allies could make a coup attempt.[76]

The socio-political system, then, as suggested in Crowley's model, is not as directly applicable to Greek siege warfare, since the evidence from *How to Survive Siege Warfare* suggests a more complex picture. The political beliefs of the defending forces were not necessarily the same even among resident combatants, and Aeneas warns of the dangers of the emergence of fifth columns as a result. In addition, where supporting troops came from other cities or from mercenary units, the political aims and objectives of the defending force could be mixed depending on which units comprised it; moreover, while there were significant cultural overlaps between groups from across the Greek world, there could also be significant differences which could lead to problems in communication and cohesion.

The Compliance Relationship

In Crowley's model, the way in which the state can force a solider to comply with the instructions given to him can be seen in three key forms: 'alienative/ coercive, calculative/remunerative and, of course, moral/normative'.[77] The first of these involves a (potentially unwilling) solider being forced to fight; the second focuses on the monetary reward a soldier receives for fighting; the third, which Crowley identifies as the most important for the Athenian hoplite: 'joins moral commitment with normative power.'[78] Crowley's model identifies the first two components of the compliance relationship as weak and flawed respectively and focuses on the third in his explanation of Athenian hoplite psychology. However, for Aeneas Tacticus, it is clear that the use (or deliberate avoidance) of the first and second elements of the compliance relationship are as important as the third, if not more so.

Aeneas advocates use of threat, and potentially actual force, to ensure the obedience and compliance of soldiers in the besieged city. The penalties he advised for disobedience, desertion, and attempting to upset the *status quo* ranged from imprisonment to capital punishment, with the perpetrator being sold into slavery as a middle option depending on the offence.[79] Unspecified penalties, in addition to a fine, should be imposed on soldiers (who may well have been mercenaries as opposed to citizen soldiers) who failed to turn up for their duty on watch.[80] The fact that Aeneas feels it necessary to advise these penalties suggests that defending forces were commonly hindered by fighters who were not completely committed to their cause, and that these units were not always as cohesive as they needed to be in a siege situation.

On the other hand, Aeneas recommends that leaders should avoid punishment of their men when they are already demoralised. For minor offences, where a commander might normally be expected to shout at his men for lack of attention or for failures in discipline, Aeneas points out that this is likely to do more harm than good and lead to the men becoming: 'more spiritless'.[81] If punishment needed to be doled out, Aeneas recommends that leaders focus on the rich, rather than the common soldiers, so as to force them to lead by example.[82] However, the list of the times when it would be appropriate to discipline the men in this way, or indeed to ignore minor offences, does not survive; this is another occasion on which Aeneas unfortunately declines to repeat his other works.[83] Aeneas, therefore, has decidedly mixed views of the efficacy of the alienative/coercive form of the compliance relationship, which reflects Crowley's findings for Athens. Athenian hoplites were subject to *graphai* which included penalties for acts of cowardice and misdemeanours on the field, which included fines and *atimia*;[84] the scale of punishments recommended by Aeneas is significantly more punitive than those found in the Athenian context, however.

When setting up his model for the compliance relationship, Crowley points out, very reasonably, that the calculative/remunerative compliance relationship has a serious flaw in that: 'soldiers tend to feel, understandably, that no amount of money is worth dying for.'[85] In *How to Survive Siege Warfare*, however, there is one key group in the defensive forces for which the calculative/remunerative compliance relationship is seen as not only very important, but which in many ways defines the group and its relationship to the city: mercenaries. Ensuring that mercenaries are paid is a crucial part of Aeneas' advice to city leaders. Without such payment, unity amongst the defenders of the city would be severely compromised. As noted above, Aeneas saw mercenary forces as a critical threat to the stability of the city which hired them and insisted on taking measures (including punitive ones) to ensure that their loyalty and discipline were maintained. The payment made to the mercenaries counterbalances the potential punishments used to keep the mercenaries in line, with rewards to induce them to provide good service to the hiring city, creating a 'carrot-and-stick' approach. According to Aeneas' advice, the best way to ensure that mercenaries were paid efficiently was to have the richest members of the city make the payments. Rather than directly reimbursing them for their costs (and therefore limiting the potential for the mercenaries to drain the city's financial resources), Aeneas suggests that the city could reduce the taxes paid by the wealthy in proportion with what they had paid out.[86] There is no evidence in *How to Survive Siege Warfare* on whether citizen soldiers were paid to carry out their duties while the city was under siege, and so it is only with regard to mercenaries that it is possible say for certain that the calculative/remunerative compliance relationship existed.[87]

The moral/normative compliance relationship is closely associated with the: 'strength of the combatant's civic identity, his belief in the effectiveness of his socio-political system, and his acceptance of its associated ideology.'[88] As a result many of the caveats given above in the section on the socio-political system must be taken into account when considering how Crowley's model can be applied to unit cohesion in *How to Survive Siege Warfare*. The likelihood that defending forces could be made up of any combination of resident or citizen fighters, mercenaries, and allied troops means that socio-political identity was not necessarily common to all of the defenders.

However, this element of the compliance relationship is clearly very important to Aeneas in maintaining cohesion among the defensive forces. The opening of *How to Survive Siege Warfare* draws on the things which Aeneas considers fundamental to the existence of a city-state: shrines, the fatherland, and family.[89] Loyalty to these three things would be key to ensuring moral commitment by every citizen soldier in the city,[90] even if they could not be called upon to keep allied and mercenary forces committed to the defence of the city. Keeping allied forces loyal was a much more difficult problem, and Aeneas recommends that these troops be supervised by loyal men from the host city.[91] Moreover, he points to the possibility of allied troops deserting, which suggests that there could only ever be a weak moral/normative compliance tie between them and the besieged city.[92]

The moral/normative compliance relationship also relies heavily on the: 'legitimacy of the demand to fight and the legitimacy of the leader making it'.[93] Again, much of this has been dealt with above: Aeneas' biggest concern during a siege is that the *status quo* be maintained,[94] which demonstrates that the perception of the city's leaders as legitimate was (or could be) precarious in the minds of the city's inhabitants. The perceived threat to the leadership of the city is tangible in *How to Survive Siege Warfare*. The moral/normative compliance relationship between the state and the soldier would therefore have significant potential to be weakened, whether or not there was a real and present challenge to the legitimacy of the city's leadership. The very suspicion that there might be a challenge to the existing regime could weaken the bond and thus make the cohesion of the units fighting to defend the city much looser. This aspect of the compliance relationship was key to a hoplite's efficacy in Crowley's model for Athenian hoplites. However, Aeneas' advice suggests that this aspect was of similar value to the alienative/coercive and calculative/remunerative compliance relationships for the effective functioning of a defensive force.

Conclusion

This study has limited itself to comparing a modern model of warfare (Crowley's *The Psychology of the Athenian Hoplite*) with an ancient one (Aeneas Tacticus' *How to Survive Siege Warfare*). In part, this has much to do with the limited evidence from ancient siege warfare on how unit cohesion was maintained in practice; evidence from historical writers fails to give us sufficient detail for the performance of the common solider, focusing instead on the actions of the elite. *How to Survive Siege Warfare*, on the other hand, gives us an intimate insight into how a leader might perceive the actions of the men beneath him and provides us with an account, if not specifically of the psychology of those who fought in sieges, then of how these men (and women)[95] should be handled in order to defend a city successfully (and within that maintain unit cohesion).

Those defending the city faced significant challenges to maintaining the cohesion of their units. The soldier's primary group might be different to the one he fought with in wars outside the city limits; even mercenary or allied units might be split up. Potentially, the fighters' bonds with each other in these units were weakened as a result. Moreover, citizen soldiers were fighting in close proximity to their family and loved ones, with the significant possibility that these traditional non-combatants could distract the attention of those defending the city. On the other hand, the presence of their parents, wives, and children could have spurred them on to fight harder on an individual basis.[96] In any case, these factors were at risk of disrupting unit cohesion within the defending force.

The socio-political system and its relationship to those fighting to defend it are also problematic. There was no guarantee that the defenders of a city shared a socio-political background, or indeed that their aims were to defend the existing regime. Allied troops and mercenaries could make up a large part of the defensive force, and there was a risk that their objectives could come to outweigh those of the inhabitants. To make matters worse, factions within the city might have the ultimate aim of overthrowing the current leaders. Mutual suspicion and the presence of outsiders were a significant risk to the maintenance of unit cohesion within the city.

Aeneas' advice is built around ideas which aim to prevent or mitigate these problems, and this is where the compliance relationship becomes a significant factor for this model. Whereas in Crowley's model the moral/normative compliance relationship is the most important for the functioning of the unit and forms the psychological imperative for the individual hoplite, the use of force and the receipt of payment for fighting were equally necessary to keep the mixed forces unified and cohesive.

It is therefore possible to use Crowley's model to analyse unit cohesion in the siege warfare of the fourth century BC, but significant adaptations must be made to account for the differences between siege warfare and fighting in the field. In particular, differences exist in the attempts during sieges to integrate allied, mercenary, and local forces in a way distinct from fighting in set piece battles. The presence of civilians and the fact that domestic life continued, even while the city was under siege, also significantly affected the behaviour of the local forces. On the whole, though, Crowley's model offers an effective framework through which to assess unit cohesion in ancient Greek siege warfare.

Notes

1 Ober (1994): 13, 17. See also van Wees (2011): 89–90.
2 Crowley (2012).

3 E.g. Whitehead (2002): 39–40; Hunter and Handford (1927): xviii; Burliga (2012): 64. For a slightly contrasting view (i.e., that Aeneas' target audience was made up of *poleis* in the Peloponnese only) see Hunter and Handford (1927): xxxi.
4 See Crowley (2012): 40–69.
5 See Crowley (2012): 80–104.
6 See Crowley (2012): 105–26.
7 E.g., Mytilene (428–427 BC, Thuc. 3.6); Leontini (427–426 BC, Thuc. 3.86); Melos (416–415 BC, Thuc. 5.114–5); Amphipolis (414 BC, Thuc. 7.9); Spiraeum (411 BC, Thuc. 8.11); Chios (411 BC, Thuc. 8.38); Chalcedon (409 BC, Xen. *Hell.* 1.3.4–7; and Diod. Sic. 13.66); and Samos (404 BC, Xen. *Hell.* 2.3.6).
8 E.g. at Plataea (429–427 BC, Thuc. 2.75, 2.77–8); Mytilene (428–427 BC, Thuc. 3.18); Scione (423 BC, Thuc. 4.131); and Syracuse (415–413 BC, Thuc. 6.99–100, 6.103, 7.2).
9 See Schofield (2014): 26–7.
10 See e.g. Thuc. 3.94; Diod. Sic. 14.14.4, 14.17.11, 14.48.5, 14.62.5, 14.90.7, 14.94.4; and Xen. *Hell.* 4.7.5, 4.8.6, 5.2.4, 5.3.3, 6.2.6, 7.2.10.
11 Aen. Tact. 11.13, 14.1–2.
12 As at Corcyra, Thuc. 3.70.
13 E.g. Aen. Tact. 10.10.
14 Aen. Tact. 10.12.
15 Although snippets emerge in some sources – for example in Aen. Tact. 3.6 and the plot of Aristophanes *Lysistrata.*
16 Hug (1877): 4–8, Hunter and Handford (1927): xi–xii, xviii and xxxiv–xxxvii, Oldfather (1923): 5 and Whitehead (2002): 8.
17 Whitehead (2002): 9. See also Hug (1877): 4–5, Oldfather (1923): 5, and Hunter and Handford (1927): xi–xii.
18 Whitehead (2002): 9.
19 Aen. Tact. 32.1–12.
20 For examples of the increasing use of catapults from the mid-fourth century onwards, see Diod. Sic. 14.50.4, 16.74.4–5, 16.75.3, 17.42.7, 17.43.1, 17.45.2; Arrian *Anab.* 1.20, 1.22, 2.23; and Curtius Rufus 4.3.13. See also Rihll (2007): 63, and Marsden (1969): 60–1 for discussion of the epigraphic evidence from this period.
21 Aen. Tact. 1.1–1.3. See also Whitehead (2002): 41. See above, note 3.
22 E.g., for a city with multiple open spaces (or only one): Aen. Tact. 2.7–8; where a city has access to horses, or terrain where cavalry or mounted messengers could be used: 6.6, 26.4; where a city has (or lacks) boats or a fleet: 16.13, 16.21; where a city has access to large numbers of wheeled vehicles: 16.15; and where a city's hinterland consists of particularly difficult terrain (or vice versa): 16.16–18.
23 Aen. Tact. 1.5–7, 5.1, and 22.15.
24 Sparta: Aen. Tact. 2.2; Plataea: 2.3–6; Chalcis on the Euripos: 4.1–4; Eleusis: 4.8–4.11; Chios: 11.3–11.6, 17.5; Argos: 11.7–10, 17.2–17.4; Heracleia: 11.10a–12, 12.5; Corcyra: 11.13–15; Chalcedon: 12.3–4; Abdera: 15.8–10; Achaia (name of city lost due to textual problems with MSS): 18.8; Teos: 18.13–19; Apollonia: 20.4; Aegina: 20.5; Naxos: 22.20; Ilion: 24.3–14, 31.24; Thrace: 27.7–10; Clazomenae: 28.5; Parion: 28.6–7; Potidaea: 31.25–7; Epeiros: 31.31–2; Lampsacus: 31.33; Barca: 37.6–7; and Sinope: 40.4–5.
25 Aen. Tact. 5.1. See also note 23 above.
26 Crowley (2012): 7.
27 Crowley (2012): 7.
28 Crowley (2012): 7–11.
29 Aen. Tact. praef.2: *hierōn kai patridos kai goneōn kai teknōn kai allōn.*
30 Aen. Tact. praef.1.
31 Aen. Tact. 5.1.

32 See above.
33 Aen. Tact. 3.6: *oikonomountes pros touskat' oikon, tekna kai gunaikas.*
34 Aen. Tact. 10.23–25.
35 Aen. Tact. 10.23.
36 Aen. Tact. 10.24–25.
37 E.g. Aen. Tact, 10.6. Prisoners held in foreign cities might also be released to cause trouble at home: see Thuc. 3.70 on the *stasis* at Corcyra having its roots in such a venture.
38 See above, note 31.
39 It is worth noting that Aeneas gives no suggestions as to how to deal with any pre-existing disputes among neighbours beyond his general comments on how to maintain unanimity in the city as a whole. Moreover, as discussed below, these units would not necessarily be the same as those the men would fight in outside the city, potentially weakening the cohesion of these neighbourhood-based fighting groups.
40 Crowley (2012): 27–35 and 43–5.
41 It is worth noting, however, that Aeneas did expect cities to be subdivided into smaller units of citizens; however, he is very careful to avoid using any term which might specify the nature of these subdivisions, e.g. tribes, Aen. Tact. 3.1–2.
42 Aen. Tact. 14.2.
43 And therefore betray the city or engage in civil disobedience, Aen. Tact. 14.1.
44 E.g. Garlan (1975): 86–9. See also van Wees (2015): 55–6; Crowley (2012): 23; Zaccarini (2015): 221.
45 E.g. van Wees (2015): 65.
46 He is also notorious for advising commanders not to let the women throw anything in case they gave themselves away: Aen. Tact. 40.4–7.
47 Aen. Tact. 1.7, 5.1.
48 Aen. Tact. 1.4.
49 Aen. Tact. 3.4: *protōn men rumēs hekastēs apodeixai rumarchēn andra ton epieikestaton te kai phronimōtaton.*
50 Garlan (1975): 93 and 95–6; Trundle (2004): 105, 110–11, 117, 146–7.
51 See e.g. Whitehead (2002): 21; Burliga (2012): 72.
52 See below under 'The compliance relationship'.
53 Aen. Tact. 13.3. See also Xen. *An.* 7.2.6 and Polyaen. 2.30.1 for examples of this in practice.
54 Aen. Tact. 10.18 refers to a mercenary camp, *xenikos stratopedon*, where *xenikos* is conventionally translated as 'mercenary' rather than 'foreigner' according to the context of the passage. Coercive relationships between cities and the mercenaries they hired are discussed in more detail below.
55 Unit cohesion within Greek mercenary units remains a much-neglected area of research, though in many ways it is likely to follow the trends examined in Crowley (2012), with the soldiers themselves loyal to their unit as a substitute for their city; see Garlan (1975): 93; McKechnie (1989): 86–8, 93; Trundle (2004): 132, 139–43, 146–7.
56 See note 4 above.
57 Crowley (2012): 81.
58 Aeneas' *exempla* cover a broad range of cities of different sizes and population levels (see note 24 above), while his more general advice deliberately avoids commenting on the geographical area available to his hypothetical *polis.*
59 Crowley (2012): 84.
60 See note 23 above.
61 See note 43 above.
62 Referring them instead to another of his works, *Procurement* – Aen. Tact. 14.2.

63 Crowley (2012): 84–5; Garlan (1975): 86–9.
64 Crowley (2012): 85.
65 Crowley (2012): 85–6. Material evidence exists for communal cult activities throughout the Greek world, including 'Messenia, the Ionian Islands, Boeotia, Laconia, and Crete', Alcock (1991): 449.
66 Aen. Tact. 17.1: *eni gar kai en toiōde kairō sphalēnai tous heterous* ('for in such a circumstance one [faction is able] to overcome the others'). See also 10.4, 17.2–6, 22.16–17, 29.3ff.
67 Aen. Tact. 22.16–17.
68 Aen. Tact. 10.4–5, 17.1–6, 22.16–18, 29.3–10.
69 Aen. Tact. 10.20–21.
70 Aen. Tact. 10.15.
71 For example, in Germany under the Nazi regime, Gellately (1996): 943, 945–9, 952. Other examples include denunciations in the USSR (see e.g., Alexopoulos (1999): 648–51) and China (see e.g., Jiaqi and Gao (1996): 256, 264).
72 Aen. Tact. 24.1: *paradidonta de sunthēmatadeipronoein, eantuxē to strateuma migades ontes apo polemeōn ē ethnōn, hopōs mē, an parechē to en eidos duo onomata, amphiboles paradothēsetai, hoion tade, Dioskouroi Tundaridai, peri henos eideos duo onomata ou ta auta* … ('When giving out passwords it is necessary to plan ahead, if the army happens to be mixed from [different] cities and tribes, so that, if two words mean the same thing, [the password] will not be given out ambiguously, such as this – Dioskouroi and Tyndaridai – two dissimilar words for the same thing …'). Other examples he gives include Ares and Enalyos, Athena and Pallas, sword and dagger, and torch and light, 24.2.
73 Aen. Tact. 24.15: *koinotata pasin.*
74 For example, when allies of the Chalcedonians refused to take part in the besieged city's commanders' plans, Aen. Tact. 12.3.
75 Aen. Tact. 10.19.
76 Aen. Tact. 12.4–5.
77 Crowley (2012): 19.
78 Crowley (2012): 20.
79 Aen. Tact. 10.19.
80 Aen. Tact. 22.29. Whitehead (2002): 161–3, suggests that this section is targeted towards cities employing mercenaries, and that the unspecified 'usual penalty' refers to that of imprisonment mentioned at 10.19.
81 Aen. Tact. 38.4: *athumoteroi gar eien an.*
82 Aen. Tact. 38.5.
83 In this case, the list would have appeared in his work on *Addresses*. Aen. Tact. 38.5.
84 Crowley (2012): 106–7.
85 Crowley (2012): 19.
86 Aen. Tact. 13.2–4.
87 This excludes the advice Aeneas gives on the cancellation of debts; he does not specify to whom such measures should apply (i.e. whether there was debt relief for everyone, or if only soldiers were eligible). See above.
88 Crowley (2012): 109.
89 Aen. Tact. praef.2.
90 Aen. Tact. praef.3.
91 Aen. Tact. 3.3.
92 Aen. Tact. 26.7.
93 Crowley (2012): 109.
94 See note 23 above.
95 Aen. Tact. 40.4–5.

96 See e.g., at Selinus in 409 BC, when old men and women supported fighters on the walls, Diod. Sic. 13.55.4, and at Himera in the same year, where Diodorus 13.60.4–5 tells us that the presence of civilians on the walls spurred on the fighters.

Bibliography

Alcock, S. E. 1991: 'Tomb Cult and the Post-Classical Polis', *American Journal of Archaeology* 95 (3): 447–467.

Alexopoulos, G. 1999: 'Victim Talk: Defense Testimony and Denunciation Under Stalin', *Law & Social Inquiry* 24 (3): 637–654.

Burliga, B. 2012: 'The Importance of the Hoplite Army in Aeneas Tacticus' *Polis*', *Electrum* 19: 61–81.

Crowley, J. 2012: *The Psychology of the Athenian Hoplite: The Culture of Combat in Classical Athens*, Cambridge.

Garlan, Y. 1975: *War in the Ancient World: A Social History*, New York.

Gellately, R. 1996: 'Denunciations in Twentieth-Century Germany: Aspects of Self-Policing in the Third Reich and the German Democratic Republic', *The Journal of Modern History* 68 (4): 931–967.

Hug, A. 1877: *Aeneas von Stymphalos*, Zurich.

Hunter, L. W. and S. A. Handford, 1927: *Aeneas on Siegecraft*, Oxford.

Illinois Greek Club (tr.), 1923: *Aeneas Tacticus, Asclepiodotus, Onasander* (Loeb Classical Library), Cambridge, Massachusetts.

Jiaqi, Y. and G. Gao, 1996: *Turbulent Decade: A History of the Cultural Revolution* (trans. and ed.D. W. Y. Kwok), Honolulu.

Marsden, E. W. 1969: *Greek and Roman Artillery: Historical Development*, Oxford.

McKechnie, P. 1989: *Outsiders in the Greek Cities in the Fourth Century BC*, London and New York.

Ober, J. 1994: 'Classical Greek Times'. In M. Howard, G. J. Andreopoulos and M. R. Shulman (eds.), *The Laws of War: Constraints on Warfare in the Western World*, New Haven and London: 12–26.

Oldfather, W. A. 1923 (reprint 2001): 'Introduction'. In *Aeneas Tacticus, Asclepiodotus, Onasander* translated by the Illinois Greek Club (Loeb Classical Library), Cambridge, Massachusetts: 1–25.

Rihll, T. 2007: *The Catapult: A History*, Yardley.

Schofield, A. 2014: *Experimental Archaeology and Siege Warfare: Analysing Ancient Sources through Experimentation*, unpublished doctoral thesis, University of Manchester.

Trundle, M. 2004: *Greek Mercenaries form the Later Archaic Period to Alexander*, London and New York.

van Wees, H. 2011: 'Defeat and Destruction: The Ethics of Ancient Greek Warfare'. In M. Linder and S. Tausend (eds.), *Böser Krieg*, Graz: 69–110.

van Wees, H. 2015 (reprint from 2004): *Greek Warfare: Myths and Realities*, London and New York.

Whitehead, D. 2002 (2nd ed.): *Aeneias the Tactician: How to Survive Under Siege*, Bristol.

Zaccarini, M. 2015: 'Thucydides' Narrative on Naval Warfare: *Epibatai*, Military Theory, Ideology'. In G. Lee, H. Whittaker, and G. Wrightson (eds.), *Ancient Warfare: Introducing Current Research, Volume 1*, Newcastle upon Tyne: 210–228.

4 'Once within the Gates': Storming Cities and Unit Cohesion in Ancient Mediterranean Warfare

Gabriel Baker

In 219 BCE, two Aetolian commanders launched a daring surprise attack against the Achaean town Aegeira. Sailing across the Gulf of Corinth by night, their troops disembarked west of the community. Meanwhile, a deserter led twenty Aetolians into the city through an aqueduct; these infiltrators killed the guards and opened the gates for their compatriots. Unopposed, the main force dashed inside the walls and marched towards the agora. The Aegeiratans were thrown into confusion, and many fled the city while others ran to the citadel. However, rather than adequately secure the town, the Aetolian invaders instead scattered to loot. As the Aetolians became progressively disorganized, many Aegeiratans regrouped at their citadel until Dorimachus, one of the Aetolian leaders, recognized the danger. He attempted to gather his troops and dislodge the defenders – but to no avail. The Aegeiratans repulsed him, counterattacked, and drove out the Aetolians in a total rout.[1]

Perhaps it was reasonable for the Aetolians to expect a quick victory. After all, Ziolkowski writes, '[a] town had to be exceptionally well defended to stand a chance of repulsing the attackers who had managed to get a foothold within the walls; usually its fate was sealed once the breach had been made'.[2] Lee notes that defenders often panicked when attackers got through their walls, and 'simply collapsed' following a successful assault.[3] To their arguments one may add that many Mediterranean cities were small, with intramural areas of ten hectares or less.[4] It was perhaps more typical for armies to sweep through enemy towns and rapidly subdue the population; hence the Aetolians' expectations at Aegeira.

But for Polybius, the Aetolians' fatal mistake was clear:

> for considering that the occupation of a foreign city is finished when one is once within the gates, they acted on this principle, so that, after keeping together for only quite a short time in the neighbourhood of the market-place, their passion for plunder caused them to disperse.[5]

And doubtlessly, he makes an important point: in ancient warfare, the capture of a city was not always assured when the walls were breached.

DOI: 10.4324/9781315171753-5

Fighting might continue within the urban enceinte, where attackers faced unfamiliar streets and possible counterattack.[6] Moreover, local resistance could (and did) become atomized into street-by-street, building-by-building combat. Lee too is probably right that 'urban combat during the capture of cities likely occurred more often than classical texts might suggest'.[7] In such cases, as Polybius shows, attacking soldiers risked a complete reversal if they broke ranks before properly securing an enemy town.

But not all soldiers were as reckless as the Aetolians at Aegeira, and many remained together 'once within the gates'. Utilizing evidence from a range of Greek and Roman texts, this chapter argues that some ancient soldiers maintained cohesion when they invaded urban communities by force or stratagem. After breaching the walls, attacking forces might stay together in formed units or ad hoc groups. Additionally, they cooperated to assault urban bastions like marketplaces and citadels, and to suppress real or potential resistance from a city's inhabitants. Lastly, although communication was difficult in cities, commanders took steps to coordinate their soldiers' actions in the heat of urban combat.

These forms of military cooperation could be reasonably described as 'task cohesion'. In modern military research, task cohesion indicates soldiers' mutual commitment to shared goals.[8] Given that ancient armies frequently sacked cities after capturing them, the desire for loot and the need to suppress threats would have offered powerful task-related goals, motivating invading armies to fight cohesively in order to secure enemy towns.[9] However, our evidence makes it difficult to examine ancient soldiers' motives, in combat or otherwise. Thus, this chapter adopts a concept of 'cohesion' that focuses less on motivation and more explicitly on the actions of soldiers in combat. Following Anthony King, I define cohesion as 'collective combat performance', i.e., soldiers' ability 'to act together and to achieve their mission' – in this case, securing enemy towns – 'in the face of the enemy'. In King's view, 'cohesion refers most accurately to collective action itself … not to the sentiments which encourage that performance'.[10]

Before proceeding to the main argument, some methodological issues deserve comment. This chapter uses evidence from Greek and Latin historical narratives, which are notorious for being formulaic, rhetorical, or dubious. Using these narratives to reconstruct ancient warfare presents special problems. Most notably, while ancient historians possessed some concept of historical truth, most did not scrupulously interrogate documents or witnesses to get to the 'facts' of military history. Even historians like Thucydides and Polybius, who seem to have placed special weight on contemporary and eyewitness evidence of war and combat, would have faced insurmountable challenges: witnesses might not remember accurate details from military engagements; and in the chaos of combat, even knowledgeable participants would have had trouble understanding what was happening beyond their immediate surroundings. Surely problems of memory, faulty witnesses, and simple confusion would have been compounded by the constrained conditions of urban warfare, where streets and buildings further limited perspectives and multiplied the bedlam.[11]

Nevertheless, our sources can still teach us a great deal about ancient warfare and urban combat. As Bosworth and Lendon argue, ancient authors seldom invented material from whole cloth.[12] And for Roth, since ancient military actions shared many basic characteristics, seemingly formulaic descriptions of military events are not necessarily fantastical imaginings of fictional battles.[13] Further, authors such as Thucydides and Xenophon also had personal military experience which shaped their depictions of warfare. Some historians, like Josephus and Polybius, had crucial first-hand experience of prolonged urban combat: Josephus watched the bitter siege of Jerusalem 'from the vantage of Roman headquarters', and Polybius was with Scipio Aemilianus during the terrible fall of Carthage.[14] Although authors like Livy lacked military experience, that was not the case for all of their audiences or for their sources.[15] Knowledgeable ancient authors and readers would have known, at least in a broad sense, what was possible and what might take place in war, including urban combat; repeated descriptions of armies fighting cohesively in cities would not have made sense to them unless they reflected known military encounters or at least seemed plausible. In short, knowledge and expectations both constrained and informed descriptions of street-fighting in our texts – so although these descriptions may be more or less 'historical' in their details, they still tell us something about how ancient armies fought in cities.

This chapter also takes a synoptic approach to the sources, examining descriptions of urban warfare from across the ancient Mediterranean world, in Greek and Roman sources spanning roughly the fifth century BCE to the first century CE. There are several justifications for this approach. First, military technology changed little in antiquity.[16] It is true that siege warfare saw technological developments in (e.g.) engineering and artillery, but the fundamental mechanics by which armies stormed and captured cities remained largely unchanged: armies entered towns by force, treachery, or stratagem; within the walls they faced streets, buildings, surviving defenders, and inhabitants.[17] Generally speaking, from there they either prevailed – wielding nothing more elaborate than fire or sword to suppress resistance – or were routed by the rallying population.[18] Due to these similarities, this chapter collectively examines many accounts written over a long period. Second, a broad view of the evidence allows patterns to emerge across authors and texts. These patterns show us how ancient writers and readers envisioned urban combat, the connotations evoked by this form of combat, and the basic forms of military cohesion that they considered plausible in these situations.[19]

Dangers Within The Gates

Ancient cities were dangerous for attacking armies, even after the collapse of the outer defences. As Lee explains, unknown streets, tight alleyways, and every building and rooftop threatened to turn a successful assault into a disastrous rout.[20] Urban inhabitants also might resist well after their walls were breached. Not only are these risks frequently stressed in our texts, but the sources are clear that the cooperative action of defenders – their continued cohesion in the face of invasion – could repulse invaders.

Several ancient sources describe defenders rallying together to drive out an enemy within their walls. As discussed above, the Aegeiratans reportedly gathered at their citadel to repel their Aetolian foes. The same risks that threatened the Aetolians are famously illustrated in the prelude to the Peloponnesian War. According to Thucydides, a Theban force infiltrated Plataea under cover of night and marshalled in the marketplace. But though the Plataeans were initially caught by surprise, they rallied, built street barricades, and fought back: their men charged the Theban troops while women and slaves pelted the invaders with stones and rooftiles. The Thebans formed up and resisted several assaults, but ultimately buckled and broke under the ferocity of the counterattack. Fleeing through unfamiliar streets in the dark and mud, many Thebans were killed or captured by their pursuers who were conversant with the terrain.[21] Diodorus conjures up a similar scene when Eumachus, a commander serving under the Syracusan tyrant Agathocles, marched into Miltinē in Libya. The inhabitants 'gathered together against him and overpowered him in the streets', driving out the foreign army.[22] Here too, the inhabitants cooperated to expel an attacking force.

Not every population could successfully repel an invading army, but many sources describe defiant defenders bogging down invaders in brutal street fighting. Thus, Diodorus recounts, the Carthaginians met tough resistance when they punched through the fortifications of Selinus. The Selinuntians, realizing that their walls were lost, fell back to barricade the streets and fight in the alleyways, as local women and children rained down missiles from the rooftops. The Carthaginians could neither surround nor come to grips with many of the defenders and were mauled in the struggle. When the Selinuntians were finally overwhelmed in the streets, they regrouped again in the agora for a last stand before organized resistance crumbled.[23] Josephus similarly describes the Roman assault on Japha, a city held by Galilean rebels: here too the inhabitants regrouped within the walls, the men fighting the Romans in alleyways while women threw down improvised missiles from the housetops. The clash reportedly lasted for hours until many defenders had been killed and the fighting devolved into a massacre.[24] And according to Plutarch, the population of Rome slowed down Sulla's army as the latter burst into the city. Throwing tiles and stones from the roofs, the inhabitants 'stopped their further progress, and crowded them back to the wall'.[25] Sulla had to attack them with firebrands and arrows until they desisted.

Other sources describe defenders digging in at citadels, marketplaces, and other strongpoints where they continued to cause trouble for their assailants. In the aforementioned Aetolian attack on Aegeira, the defenders not only regrouped at their citadel, but launched a successful counteroffensive.[26] Even though the citadel was unwalled, Polybius explains, the elevated terrain afforded an advantage against the invaders.[27] Less successfully, the Thebans made a last stand in their agora against Macedonian-led attackers before their defence melted into a rout.[28] Presumably they chose the agora because the open space allowed more room to fight.[29] We see similar behaviour in other anecdotes. Plutarch and Diodorus, describing the internecine strife in Syracuse in 357 BCE, say that Dion's mercenaries pushed into the city and drove back the soldiers of Dionysius II. But when the defenders retreated to the citadel, Dion was forced to besiege them.[30] Similarly, Polybius and Livy relate that when Scipio Africanus assaulted New Carthage, the Carthaginian commander initially resisted from the citadel.[31]

The importance of strategic urban terrain is also emphasized in ancient military manuals, in their efforts to provide useful advice to the actual practitioners of warfare. In his fourth century BCE treatise *How to Survive Under Siege*, Aeneas Tacticus advises defenders to occupy marketplaces, theatres, and other open locations, as well as municipal buildings and 'the strongest place in the whole city and the most conspicuous'.[32] The author's main concern is to prevent uprisings within besieged towns, but he obviously highlights the strategic value of these places. Indeed, he tells defenders to occupy places that are 'defensible' and well positioned for controlling a given city.[33] Additionally, in the *Strategikos*, Onasander warns that citadels can be used for counterattacks and allow defenders to maintain a foothold in an otherwise-captured city;[34] as he puts it, many defenders have 'crowded into the fortified citadel from which they have caused great labour and loss to their adversaries, who must enter into a second … and longer siege, one that is sometimes more distressing and attended by great hardships'.[35] For these authors, strategic terrain allowed defenders to hold their home turf more successfully against threats from within or without.

In sum, even after the fall of the outer defences, urban defenders could conceivably fight on in streets and marketplaces, from behind barriers or atop homes. The whole population might participate in this defence, with women, slaves, or even children fighting alongside the male inhabitants.[36] As Polybius points out, dispersal or a lack of cooperation in this environment could be disastrous, allowing cohesive defenders to rout the initially successful attackers. And it was not just looters who were vulnerable. Thucydides shows why it was important as a general rule for soldiers to maintain cohesion, even in seemingly conquered cities. In his narrative, the Theban force at Plataea managed to repulse 'two or three' attacks by the rallying Plataeans while they stayed in their ranks; yet when the Theban troops panicked and broke formation, they were easily overrun as they scattered through the streets.[37]

Interestingly, a nineteenth-century manual on urban warfare warns against many of the dangers narrated in ancient sources. Written by the French Marshal Bugeaud in 1847 and drawing upon his own experiences, this manual was meant to 'show how troops could ruthlessly put down insurrections in

Algiers'.[38] Bugeaud contends that defenders fighting in city streets can be difficult to overcome, and stresses the attackers' need for boldness, rapidity, and caution. He insists that bricks, tiles, and other projectiles thrown from windows and balconies were particularly dangerous – even more dangerous than musket fire.[39] He explicitly advises soldiers to maintain small-unit cohesion, operating in small detachments of 20 to 40 men, plus an officer, to clear out street barricades and defended buildings. Additionally, if attackers are forced to enter buildings, organization and discipline are required; otherwise, confusion results and isolated soldiers may be threatened.[40] One should not send men to attack houses without good reason, he asserts, since doing so can progressively reduce the number of troops operating in formation.[41] Put another way, avoid dispersion and remain in groups. While military circumstances and historical contexts were plainly different, Bugeaud's warnings closely echo the dangers described in Greek and Roman texts, strongly suggesting that these were real concerns in the ancient Mediterranean world as well, rather than the dramatic license of Greek and Roman authors.

Collective Action Within the Gates

Ancient assault troops, confronted with an astonishing array of threats within hostile cities, must have felt enormous pressure to stick together. Indeed, modern research suggests that shared danger encourages cohesion in military forces, fostering cooperation between soldiers as they seek to achieve their goals or counter threats.[42] It is perhaps unsurprising, then, that several Greek and Latin authors describe soldiers fighting together after they had pushed inside enemy towns. In some sources, attackers stay in units or groups as they move through the streets and suppress lingering threats; in others, they focus their efforts against urban redoubts and strategic terrain. These cooperative actions demonstrate cohesion, or King's 'collective combat performance'. For King, 'cohesion is demonstrated when soldiers are able to shoot, move, and seek cover together or in mutually supportive ways', or when their activities are successfully coordinated towards the same objectives.[43] While his concept of cohesion stems from his work on modern armies, the principles apply to ancient soldiers who fought together in hostile urban terrain.

Several ancient texts depict soldiers operating cohesively in units or groups while fighting inside enemy towns. Livy and Polybius, for example, indicate that some of Scipio's soldiers stayed in battle order as they stormed New Carthage. Livy writes that some Romans entered the city in 'regular formation... with its officers' (*iusta acies cum ducibus*); and Polybius says that Scipio commanded a group of about a thousand men inside the city.[44] Livy similarly has L. Scipio's troops entering the gates of Orongis, an Iberian city, formed up with their standards (*signa ... inlata*).[45] Describing Sulla's push into Rome in 88 BCE, Appian and Plutarch report that Sulla sent subordinate commanders with bodies of men to occupy gates and bridges, others to stand guard outside the walls, and still others to drive towards the urban centre.[46] In Orosius' and Florus' version, Sulla leads 'columns' (*agmena*) of troops through the city.[47]

Plutarch and Diodorus describe similar collective action during the fighting in Syracuse in 357 BCE. Once the mercenaries of Dionysius II had breached the walls, Diodorus writes, they were able to overpower the Syracusans and capture parts of the city because they 'numbered more than ten thousand and their lines were so well marshalled that no one was able to withstand their sheer weight'.[48] According to Plutarch, Dion's troops tried to wrest Syracuse from Dionysius' mercenaries and maintained cohesion as they fought through the city. Once inside, Dion '[divided] his commands and [formed] companies in column, [so] that he might make a more formidable attack from many points at once'.[49] These units marched through streets choked with corpses and smoke, trying 'to keep together and not break their ranks' – despite coping with difficult terrain and a narrow front.[50]

In all of these cases, the attackers' continued collective action was apparently a response to ongoing resistance within the walls. At New Carthage, the Carthaginians still held the citadel and another large hill; the population itself, moreover, had been armed and actively participated in the city's defence. Though the townspeople scattered when the Romans breached the walls, they could conceivably regroup and counterattack.[51] For his part, Sulla faced resistance from Roman inhabitants in their homes, and from Marians who had mustered in the Esquiline forum.[52] And groups of Syracusans continually 'formed to meet [Dionysius' mercenaries] in narrow alleys and other streets',[53] while Dion's counterattack had to face down those same mercenaries in the streets and at the citadel.[54] In each case, the cohesion of attackers allowed them to overpower resistance, ultimately forcing the opposition to surrender or flee, when they could be 'pursued and slain' by the more cohesive attackers.[55]

Many texts also describe attackers moving against strongpoints within cities, such as citadels and marketplaces. This is especially powerful evidence of cohesion, since it is unlikely that scattered or looting soldiers would be able to capture these places on their own. For example, several sources depict invaders marching first to marketplaces, securing this significant location before breaking off to plunder. When Dionysius II's mercenaries attacked Syracuse, they first captured the agora; only then did they move off to pillage. As noted previously, the Aetolians moved towards the agora at Aegeira before scattering to loot, just as the Carthaginians routed the last defenders in the Selinuntian agora before sacking the city.[56]

In other cases, invaders broke through the walls of a city, then launched assaults against several key positions within. Once they had breached the walls of New Carthage, some of Scipio's legionaries attacked and captured a hill garrisoned by Carthaginian troops; at the same time, Scipio led some men against the citadel where he fought with the garrison commander.[57] Similarly, Megaran colonists captured Sicilian Leontini by seizing the citadel and marketplace.[58] And as noted above, Sulla reportedly sent subordinate commanders towards specific strongpoints in Rome – the Esquiline and Colline gates and Pons Sublicius – while he led others towards the city centre.[59] Orosius adds that Sulla led a column of troops to the Forum, while Florus has him capturing the Capitoline.[60] Finally, in the climaxes of the Third Punic War and First Roman-Jewish War, the Romans had to capture multiple urban redoubts before capturing a final stronghold. Thus Appian describes the Romans pressing determinedly towards Byrsa hill, the citadel of Carthage. In his vivid retelling, the Romans had to ascend three different streets, each flanked with six-storey buildings from which the population desperately resisted. In order to make headway, the Romans were forced to clear out each building one by one, expelling or killing the inhabitants in each; they also had to fight for the streets themselves, where they met equally stiff resistance.[61] Josephus similarly has the Romans fighting through successive layers of defence in Jerusalem, facing ambushes and counterattacks in the streets, before finally assaulting the Temple where the Jewish fighters made their last stand.[62] In all of these cases, the invaders are quite clearly working together to capture defended positions.

Our texts describe other, less obvious forms of collective action when attacking armies secured hostile cities. As demonstrated above, the whole population could pose an active threat to invaders. Consequently, numerous sources describe armies invading towns and killing the inhabitants. For example, Livy, Diodorus, Thucydides, and Sallust describe Roman and Greek armies storming cities and killing the adult males.[63] Other authors describe soldiers entering cities and killing the inhabitants indiscriminately.[64] During the capture of Thebes, for example, the Macedonians' Greek allies reportedly slew the Thebans where they found them, whether in their homes or at their shrines, whether resisting or not.[65] Diodorus, describing the same event, writes that the attackers cut down 'all whom they met without sparing any'.[66]

These ruthless massacres might stem from a number of factors, not least the rage and frustration of vengeful troops who suffered through a difficult siege or costly assault.[67] But several modern scholars argue that these mass killings served a pragmatic, military function, a claim that is supported by some ancient sources.[68] In Polybius' view, the Romans indiscriminately massacred enemy populations to terrorize survivors into surrender.[69] At New Carthage, for instance, the Carthaginian commander had armed 2,000 of the inhabitants and ordered the rest to defend the walls. When Scipio commanded his legionaries to 'to kill all they encountered, sparing none', the inhabitants were not neutral bystanders; rather, they were proven enemies who remained threatening as long as they might recover or counterattack.[70] According to Livy, Hannibal ordered his army to kill the adult males in Saguntum as a matter of military necessity, since he believed that the Saguntines would resist to the last.[71] In other cases, ancient armies reportedly engaged in mass killing *before* they moved off for plunder, suggesting that they did not seek loot until they had satisfactorily suppressed local resistance. When Titus' soldiers finally breached the upper city in Jerusalem, Josephus writes, they were puzzled at the absence of defenders. Yet they still '[poured] into the alleys, sword in hand', and 'massacred indiscriminately all whom they met'.[72] They only sought plunder after taking these violent measures, perhaps unwilling to take any risks.[73]

Given their military function, these mass killings too can be regarded as a kind of collective combat action – and a few sources even describe organized groups of soldiers killing disorganized inhabitants. In the climax of the First Roman-Jewish War, Josephus says the Roman legions were 'marching all together' (*sympheroménōn*) as they moved against the inner temple precincts, killing those they met and suppressing the last pockets of resistance.[74] Polybius writes that groups of Carthaginians and Celts cut down the Romans they found in the streets of Tarentum; their targets were 'in disorder and in scattered groups' (*atáktōs kaì sporádēn*).[75] Arrian pointedly mentions massacre victims *lacking* cohesion, writing that the Thebans were 'no longer in any sort of order' (*éti oudenì kósmōi*) when Macedonians and Boeotians overran Thebes; the Thebans' failure to offer collective resistance allowed their foes to put them to the sword wherever they found them.[76] In these passages, at least, it seems that attackers were grouped together as made their way through the streets and killed the inhabitants.[77] Finally, as discussed in the next section, these massacres often stemmed from the direct orders of commanders, again indicating collective and coordinated action.

To conclude this section, many Greek and Roman sources describe soldiers bursting into enemy towns and fighting cohesively to crush local resistance. Ancient authors may have taken dramatic liberties with some of the individual anecdotes, or simply described plausible scenarios when they or their sources could not make sense of a street battle.[78] However, what is important is less the historicity of the individual episodes than the larger patterns that emerge from them, which must have made sense to knowledgeable writers and readers. And they show, first of all, that invaders might move through hostile streets in formations or organized groups; second, that attackers might march against specific locations or buildings, usually strategic terrain or hotbeds of continued resistance; and third, that soldiers might stick together as they ruthlessly massacred a city's inhabitants. Each of these activities implies a measure of cohesion, following King's understanding of military cohesion as collective action in combat: soldiers who stayed in groups and attacked urban redoubts together were manifestly coordinating their actions on the battlefield, and '[acting] together … to achieve their mission in the face of the enemy'.

Commanders, Collective Action, and Vertical Cohesion Within the Gates

As demonstrated above, armies did not always devolve into frenzied looters when they stormed cities, particularly when they sought to deal with real or imagined threats within the walls. But that does not mean that attacking forces were models of discipline in the tense moments after the breach. In fact, many Greek and Latin narratives describe armies entering enemy towns by assault or stratagem, and then – like the Aetolians at Aegeira – dispersing to loot, rape, and kill.[79] Other ancient sources explicitly describe commanders' difficulty controlling their troops during urban assaults.[80] And according to some scholars, disorganization and indiscipline were the norm when ancient armies entered cities. Ziolkowski and Levithan argue that Roman commanders had little-to-no control over their troops once the latter entered enemy towns. In the aftermath of a successful assault, they contend, Roman soldiers generally massacred the male inhabitants at hand, then rapidly scattered off to ravage and plunder on their own initiative.[81] Similarly, Kern remarks that it was 'not uncommon' for generals to lose control of their troops during the sack of cities; for Gilliver as well, soldiers would often find themselves 'concealed from the eye of authority' when sacking cities.[82] Lee argues that Greek armies were capable of some coordinated action on urban terrain. However, he adds, cities created acute discipline problems, and chaotic urban combat led to a 'lack of communication and control'.[83]

While urban terrain presented special challenges for command and control, examples of indiscipline and disorder must be balanced against instances of cohesion from above. As King notes, 'the successful coordination of the activities of group members to the same goal' points towards successful military cohesion. And in many ancient sources, commanders coordinate their troops within enemy towns, facilitating their collective action in hostile urban environments.[84] This command-oriented cohesion may also be termed 'vertical cohesion', the up-and-down connections between leaders and followers. In modern research, vertical cohesion is partly concerned with motivation and interpersonal elements, such as trust between leaders and followers. But vertical cohesion also concerns leaders' ability to direct the behaviour of subordinates – which is more visible in the ancient evidence and will be the focus here.[85]

Ancient commanders fostered cohesion in the breach by directing their troops towards specific goals inside enemy cities. We have already seen that Sulla and his sub-commanders reportedly led detachments against different objectives inside Rome, that Dion arranged divisions of his soldiers inside Syracuse, and that Scipio personally led a thousand soldiers against the citadel in New Carthage. Similarly, according to Polybius, Hannibal carefully arranged his men once they had infiltrated Tarentum: he pulled out 2,000 Gauls from his main force, divided them into three groups, then sent them with officers and Tarentine collaborators to occupy the approaches to the agora. Further, Polybius adds, 'he ordered the Carthaginian and Celtic officers to put all Romans they met to the sword. The different bodies hereupon separated and began to execute his orders'.[86]

Other sources describe commanders passing orders via trumpet, messenger, or some unstated method, in order to coordinate their troops towards specific objectives. For example, Livy says that L. Scipio's cavalry dashed to capture the forum of Orongis, 'for such were their orders' (*ita enim praeceptum erat*).[87] According to Plutarch, Pyrrhus led some of his men inside Argos, then sent orders to the rest of his army via messenger.[88] And with some frequency, our sources describe Roman, Carthaginian, and Macedonian commanders signalling their troops to massacre the inhabitants of hostile towns, either by using trumpet signals or some unspecified method.[89]

Leader-driven cohesion is particularly apparent in Josephus' account of the siege of Jerusalem and Appian's (Polybian) narrative of the siege of Carthage, both of which stemmed from first-person experience of these extraordinary events.[90] As discussed above, the Romans had to fight through the streets and buildings of Carthage to reach the citadel. Scipio Aemilianus reportedly ordered his troops to clear the avenues around the citadel and personally directed the progress, 'hurrying here and there', then taking position 'on a high place where he could overlook the work'.[91] According to Josephus, Titus also directed operations while his men drove deeper into Jerusalem. After the Romans had broken through the first and second wall, Titus could not bring his whole army against the Temple because the approach was too narrow. So, he assigned detachments of a thousand soldiers to each of his tribunes, putting these under the command of a legate with orders to assault the temple by night. Meanwhile, he took up a position in the nearby fortress, Antonia, where he could observe.[92] In both these cases, the commanders were coordinating 'the activities of group members to the same goal' and 'in mutually supportive ways'; in other words, they were facilitating the cohesion of their forces.

Conclusion

Many cities were effectively captured after the walls were breached, yet invaders could also face continuing dangers within. Inhabitant populations could threaten assault troops and defenders might fight in the streets, from roofs, or fortify themselves in redoubts. Polybius says that some attackers failed to take adequate precautions – and paid the price. Others, however, continued to fight cohesively in urban terrain. This chapter has tried to show the basic ways that ancient armies maintained cohesion as they secured hostile urban environments.[93] Attackers might remain in formed units or ad hoc groups as they cut their way through a city's streets or ruthlessly killed its inhabitants; others might march together against key terrain like citadels and marketplaces. Commanders attempted to coordinate their men in these efforts, leading them personally, passing orders by signal or messenger, or dispatching detachments under the command of subordinate officers.

These conclusions must be treated cautiously, given the nature of the evidence, and there is no way to tell how frequently ancient armies employed cohesive tactics in urban combat. Still, it is worth noting that similar cohesive actions are reported in other historical periods, when medieval or early modern armies had to fight their way through hostile towns. For the sake of example, in the First Battle of St. Albans (1455), the Siege of Tenochtitlan (1521), during a rebellion against the Ghibellines in Florence (1266), and in the two sieges of Saragossa (1808–1809), contemporary or near-contemporary accounts plainly describe soldiers operating together in urban terrain, led by their commanders and officers, and coordinating their efforts against entrenched opposition.[94] The comparisons only go so far, given sharp differences in historical context and military circumstances. Yet like Bugeaud's warning about the hazards of city combat, they lend credence to the conditions described in Greek and Roman narratives. And for ancient authors, military cohesion was evidently possible – at times even necessary – once within the gates.

Notes

1 Polyb. 4.57–58.
2 Ziolkowski (1993): 79.
3 Lee (2010): 141, 152.
4 E.g. Hansen (2006): 100–6; Lee (2010): 143.
5 Polyb. 4.57.10–58.1; cf. 10.17.5. All translations are from the Loeb editions.
6 Lee (2010) discusses the dangers and confusion of urban combat. These dangers are also examined below.
7 Lee (2010): 141.
8 MacCoun and Hix (2010): 139; King (2013): 34–6.
9 For the commonality of the sack, see Ziolkowski (1993) and Levithan (2013): 205–28 (discussing Roman warfare, but their basic points hold for ancient warfare generally).
10 King (2013): 36–7.
11 Difficulties using literary sources for military history: e.g., Whitby (2008); Whatley (1964).
12 Lendon (2009); Bosworth (2003).
13 Roth (2006): 57.
14 Josephus: Goodman (2019): 3–4; Polybius: Walbank (1957): 5.
15 Livy's 'war-hardened' audience, Ziolkowski (1993): 72.
16 Lendon (2005): 8–13.
17 Levithan (2013): 13–14.
18 Kern (1999): 352–5, states that siege warfare 'remained much the same' into the Middle Ages and beyond.
19 Levithan (2013): 54: 'Given that narratives often presume a broader familiarity with the principles and customs of Roman warfare than we now possess, it behoves us to assemble all of the evidence we have in order to help us read between the lines'; cf. Ziolkowski (1993): 72.
20 Lee (2010): 143–9; Barry (1996): 55–74.
21 Thuc. 2.2.4–2.4.8.
22 Diod. Sic. 20.58.2.
23 Diod. Sic. 13.56.6–8.
24 Jos. *BJ* 3.302–304.
25 Plut. *Sull.* 9.6; cf. App. *BC* 58; Livy 5.21.10 (Veii).
26 Other examples of defenders holding out in their citadel after the rest of the town was captured: Hdt. 5.100; Polyb. 7.18.7; Livy 24.2–3; 31.45.5–6; Plut. *Sull.* 14.7.

27 Polyb. 4.58.8.
28 Arr. *Anab.* 1.8.6–8.
29 Cf. Plut. *Pyrr.* 33.2: Pyrrhus chose to fight in Argos' agora because it 'afforded him room for withdrawing and fighting'.
30 Plut. *Dion.* 46.4, 48.1; Diod. Sic. 16.20.5.
31 Polyb. 10.15.7; Livy 26.46.9. Other examples from the Second Punic War: Polyb. 8.31–34; 8.37; Livy 25.24–30; Plut. *Marc.* 18.
32 Aen. Tact. 1.9, 3.5, 22.2–4, discussed at Lee (2010): 154.
33 Aen. Tact. 1.9.
34 Lee (2010): 144.
35 Onas. *Strat.* 42.19; cf. Polyaen. 7.6.10.
36 Chaniotis (2005): 107–9; Loman (2004): 41–2; Barry (1996): 55–74.
37 Thuc. 2.4.1.
38 Graham (2004): 36.
39 Bugeaud (1997): 132.
40 Bugeaud (1997): 131, 139.
41 Bugeaud (1997): 135.
42 Shared danger and cohesion: MacCoun and Hix (2010): 146–9, with sources; Belenky and Kaufman (1983).
43 King (2013): 36–8.
44 Livy 26.46.7; Polyb. 10.15.7.
45 Livy 28.3.13.
46 App. *BC* 58; Plut. *Sull.* 9.5–7.
47 Oros. 5.19; Flor. 2.9.6. cf. Xen. *Hell.* 2.4.10–11; Caes. *BCiv.* 3.111.
48 Diod. Sic. 16.19.3.
49 Plut. *Dion* 45.3; Lee (2010): 154.
50 Plut. *Dion* 46.3.
51 Polyb. 10.12.1–3; 10.15.3–7; Livy 26.44.2, 26.45.8–9.
52 App. *BC* 58; Plut. *Sull.* 9.6–7.
53 Diod. Sic. 16.19.4.
54 Plut. *Dion* 46.3–5.
55 Plut. *Dion* 46.6.
56 Diod. Sic. 16.19.4 (Syracuse); Diod. Sic. 13.57.2 (Selinus); cf. Livy 28.3.13.
57 Polyb. 10.15.7; Livy 26.46.7–9.
58 Polyaen. 5.5.1.
59 App. *BC* 58; Plut. *Sull.* 9.5–7
60 Oros. 5.19; Flor. 2.9.6–7.
61 App. *Pun.* 128.
62 Jos. *BJ* 5.303–6.408.
63 E.g. Diod. Sic. 18.22.2; Thuc. 5.83.2; Livy 9.31.3; 26.46.10; 31.27.4; Sall. *Iug.* 91.5–7.
64 Thuc. 7.29.4; Polyb. 10.15.4–5; 8.30.4, 9; 7.18.9; Diod. Sic. 13.57.2;13.62.3–4; 14.53.1–2; Arr. *Anab.* 1.8.8; 6.11.1; App. *Hisp.* 32; Jos. *BJ* 6.271; Caes. *BG* 7.28.3; Livy 28.20.6; 42.63.10; Zon. 8.11; Paus. 7.16.8.
65 Arr. *Anab.* 1.8.8. See also Arr. *Anab.* 6.11.1.
66 Diod. Sic. 17.13.1.
67 See Levithan (2013): 218–27; Ziolkowski (1993): 83–6. For soldiers killing in anger, see e.g., Livy 28.20.6-7; Arr. *Anab.* 1.8.8; Caes. *BG* 7.28.3; cf. Reisdoerfer (2007): 70–5.
68 For Gaca (2010): 136, indiscriminate killing in Greek warfare was a way to 'stun the remaining survivors into submission'. Ziolkowski (1993): 76–7, and Levithan (2013): 208–9, 214, contend that the Romans massacred city populations in the final phase of an assault, to 'mop up' the enemy in the dangerous moments after the breach.

69 Polyb. 10.15.5.
70 Polyb. 10.15.4. See also Polyb. 10.12.2–3; Livy 26.44.2. Likewise, Goldsworthy (2000): 275: '[T]he citizens of New Carthage were not peaceful neutrals, but active participants in the defence'.
71 Livy 21.14.4.
72 Jos. *BJ* 6.404.
73 Other examples at Ziolkowski (1993): 77.
74 Jos. *BJ* 6.273.
75 Polyb. 8.30.9.
76 Arr. *Anab.* 1.8.8.
77 In some cases, soldiers scattered off and murdered as individuals rather than in cohesive groups: e.g., Livy 25.31.9; Plut. *Marc.* 19.4–5. cf. Ziolkowski (1993): 69–91.
78 Cf. Evans (2016): 178. Writing of the street fighting in Syracuse in 357, Evans asserts that '[t]he details of the battles in the streets of the city were invented to illustrate the changing fortunes of each side and that it was the Syracusans in the end who won the upper hand in the conflict'. Regardless, Diodorus and Plutarch clearly thought such details were plausible enough.
79 E.g. Thuc. 7.29.4 (Mycalessus); Diod. Sic. 13.62.3–4 (Himera); Diod. Sic. 17.13.1–5 (Thebes); Polyb. 7.18.9 (Sardis); Plut. *Sull.* 14.3 (Athens); Memn. 35.5 (Heracleia Pontica); Tac. *Hist.* 3.33.1–3 (Cremona).
80 See esp. Caes. *BCiv.* 2.12: Caesar writes that if the defences failed at Massilia (during the siege in 49 BC), 'soldiers could not be withheld from bursting into the town in hope of plunder and utterly destroying it'. See also Diod. Sic. 14.53.1–2; Caes. *BG* 7.47; Livy 37.32.11; App. *Pun.* 15. These and other episodes are discussed by Levithan (2013): 221–2.
81 Ziolkowski (1993): 69–91; Levithan (2013): 205–28.
82 Kern (1999): 79; Gilliver (1996): 226.
83 Lee (2010): 148, 153.
84 King (2013): 36–8. See also 206–7: in early twentieth-century citizen armies, King argues, unit cohesion was heavily dependent upon the presence of combat leaders who 'were not merely supposed to command their troops, coordinating their movements in combat, but to act as moral exemplars for them ... The moral authority of leadership depended on the willingness of a commander to lead [an assault] himself'.
85 MacCoun and Hix (2010): 139–40, 155–6; Salo (2011): 30.
86 Polyb. 8.30.1–4. See also Hall and Rawlings, this volume, p. 91, 96-7. Roman and Hellenistic armies had extensive command structures which could facilitate this kind of coordination, see Isaac (1998): 389–90; Hoyos (2007): 68–70, 75–6; Sekunda (2010): 460–1. But some ancient armies, like most in Archaic and Classical Greece, lacked detailed command hierarchies: van Wees (2004); Lee (2010): 150.
87 Livy 28.3.13.
88 Plut. *Pyrr.* 33.1–2.
89 E.g. App. *Hisp.* 52, *Mith.* 38; Tac. *Ann.* 12.17; Livy 21.14.4; Curt. 4.4.14; Arr. *Anab.* 4.2.4. Southern (2006): 159, shows that the Roman army used a variety of instruments for tactical signals. See also Polyb. 10.15.4, 8; Ziolkowski (1993): 78–9, 82–8.
90 For Appian's use of Polybius for the Third Punic war, see e.g. Baronowski (1995): 18 n. 8, with sources.
91 App. *Pun.* 128, 130.
92 Jos. *BJ* 6.129–135.
93 Specific armies and generals might have had unique methods to cope with these dangers, but such is beyond the scope of this study.
94 Goodman (1990): 24; Diaz del Castillo (2008): 256, 283–4, 294–5; Villani *Cronica* VII:14; Fremont-Barnes (2019): 2–17.

Bibliography

Baronowski, D. W. 1995: 'Polybius on the Causes of the Third Punic War', *Classical Philology* 90: 16–31.
Barry, W. D. 1996: 'Roof Tiles and Urban Violence in the Ancient World', *Greek, Roman, and Byzantine Studies* 37: 55–74.
Belenky, G. L., and L. W. Kaufman, 1983: 'Cohesion and Rigorous Training: Observations of the Air Assault School', *Military Review* 63: 24–34.
Bosworth, B. 2003: 'Plus ça change … Ancient Historians and Their Sources', *Classical Antiquity* 22: 167–198.
Bugeaud, T. R. 1997: *La Guerre des Rues et des Maisons du Maréchal Bugeaud* (ed. M. Bouyssy), Paris.
Chaniotis, A. 2005: *War in the Hellenistic World*, Malden.
Díaz del Castillo, B. 2008: *The History of the Conquest of New Spain* (trans and ed. D. Carrasco), Albuquerque.
Evans, R. 2016: *Ancient Syracuse: From Foundation to Fourth Century Collapse*, London.
Fremont-Barnes, G. 2019: 'Nineteenth-Century Precedents'. In G. Fremont-Barnes (ed.), *A History of Modern Urban Operations*, Cham: 1–33.
Gaca, K. 2010: 'The Andrapodizing of War Captives in Greek Historical Memory', *Transactions of the American Philological Association* 140: 117–161.
Gilliver, C. M. 1996: 'The Roman Army and Morality in War'. In A. B. Lloyd (ed.), *Battle in Antiquity*, London: 219–238.
Goldsworthy, A. K. 2000: *The Fall of Carthage*, London.
Goodman, A. 1990: *The Wars of the Roses*, London.
Goodman, M. 2019: *Josephus's* The Jewish War*: A Biography*, Princeton.
Graham, S. 2004: 'Cities as Strategic Sites: Place Annihilation and Urban Geopolitics'. In S. Graham (ed.), *Cities, War, and Terrorism: Towards and Urban Geopolitics*, Malden: 31–53.
Hansen, M. H. 2006: *The Shotgun Method: The Demography of the Ancient Greek City- State Culture*, Columbia.
Hoyos, B. D. 2007: 'The Age of Overseas Expansion'. In P. Erdkamp (ed.), *Companion to the Roman Army*, Malden: 63–79.
Isaac, B. 1998: *The Near East under Roman Rule: Selected Papers by Benjamin Isaac*, Leiden.
Kern, P. B. 1999: *Ancient Siege Warfare*, Bloomington.
King, A. 2013: *The Combat Soldier: Infantry Tactics and Cohesion in the Twentieth and Twenty-First Centuries*, Oxford.
Lee, J. W. I. 2010: 'Urban Warfare in the Classical Greek World'. In V. D. Hanson (ed.), *Makers of Ancient Strategy: From the Persian Wars to the Fall of Rome*, Princeton: 138–162.
Lendon, J. E. 2005: *Soldiers and Ghosts: A History of Battle in Classical Antiquity*, New Haven.
Lendon, J. E. 2009: 'Historians without History: Against Roman Historiography'. In A. Feldherr (ed.), *The Cambridge Companion to the Roman Historians*, Cambridge: 41–61.
Levithan, J. 2013: *Roman Siege Warfare*, Ann Arbor.
Loman, P. 2004: 'No Women, No War: Women's Participation in Ancient Greek Warfare', *Greece & Rome* 51: 34–54.

MacCoun, R. J. and W. M. Hix, 2010: 'Unit Cohesion in Military Performance'. In B. D. Rostker *et al.* (eds.), *Sexual Orientation and U.S. Military Personnel Policy: An Update of RAND's 1993 Study*, Santa Monica: 137–165.
Paul, G. M. 1982: '"Urbs Capta": Sketch of an Ancient Literary Motif', *Phoenix* 36: 144–155.
Reisdoerfer, J. 2007: '... non aetate confectis, non mulieribus, non infantibus pepercerunt: Etude sur le massacre d'Avaricum (BG VII 28)', *Göttinger Forum für Altertumswissenschaft* 10: 59–80.
Roth, J. 2006: 'Siege Narrative in Livy: Representation and Reality'. In S. Dillon and K. Welch (eds.), *Representations of War in Ancient Rome*, New York: 49–67.
Salo, M. 2011: *United We Stand – Divided We Fall: A Standard Model of Unit Cohesion*, Helsinki.
Sekunda, N. 2010: 'The Macedonian Army'. In J. Roisman and I. Worthington (eds.), *Companion to Ancient Macedonia*, Malden: 446–471.
Southern, P. 2006: *The Roman Army: A Social and Institutional History*, Oxford.
van Wees, H. 2004: *Greek Warfare: Myths and Realities*, London.
Villani, G. 1907: *Villani's Chronicle* (trans. R. E. Selfe, ed. P. H. Wicksteed), New York.
Walbank, F. W. 1957: *A Historical Commentary on Polybius*, vol. 1, Oxford.
Whatley, N. 1964: 'On the Possibility of Reconstructing Marathon and Other Ancient Battles', *Journal of Hellenic Studies* 84: 119–139.
Whitby, M. 2008: 'Reconstructing Ancient Warfare'. In P. A. G. Sabin, H. van Wees, and M. Whitby (eds.), *The Cambridge History of Greek and Roman Warfare, Volume 1: Greece, the Hellenistic World, and Rome*, Cambridge: 54–84.
Ziolkowski, A. 1993: '*Urbs direpta*, or How the Romans Sacked Cities'. In J. Rich and G. Shipley (eds.), *War and Society in the Roman World*, New York: 69–91.

5 Unit Cohesion in the Multi-Ethnic Armies of Carthage

Joshua R. Hall and Louis Rawlings

Carthaginian armies and expeditionary forces consisted of ethnically diverse soldiers and warriors from around the central Mediterranean. Ancient sources gloss these constituent parts with broad ethnic terms, such as Celts, Iberians, Africans, Ligurians, Libyans, Numidians, Phoenicians, Liby-Phoenicians, and Carthaginians. Many of these descriptors were used carelessly and imprecisely, reflecting at times the discourse of 'otherness' present in ancient historical narrative. Rarely are specific communities, tribes, or leaders mentioned. This leaves much to be desired from the perspective of the modern scholar of Punic warfare. For instance, the subtleties of Iberian tribal identities, or Etruscan city origins, are not typically conveyed by our sources. Of course, it is possible that these things mattered little to Punic recruiters either.

Nevertheless, many of the armies that Carthage fielded throughout its history contained within them a morass of individual and group identities, rivalries, differences in language, cultural preconceptions, and other factors which we can barely discern. All of these issues, though, Punic commanders had to come to grips with and mould the various actors and groups into coherent and effective armies. Commentators, ancient and more recent, have regarded this task as difficult and sometimes un-accomplished. Polybius, in his constitutional digression, criticised the structure of Carthaginian military forces, declaring them to be inferior to their Roman counterparts because of this multifaceted composition (Polyb. 1.67.3, 7, 6.52).[1] For the Renaissance polymath Niccolò Machiavelli, Punic armies were prime examples of the inferiority which came from multi-national, non-citizen forces.[2] While these jejune criticisms are understandable in context, they are almost certainly constrained by the genre and rhetorical goals of the authors.

The reality of how well the armies of Carthage functioned in the field is much more complex than any gloss could allow. We must be sensitive to their objectives, which varied from campaign to campaign, and ranged from projection of power to tactical superiority. One of the most effective and important means of assessing their functionality is by analysing unit cohesion therein. Modern research on the phenomenon has shown that the ancient, and later, preoccupation with the supposed faults of Punic forces may be inconsequential to the multivariate ways in which they were made

DOI: 10.4324/9781315171753-6

structurally and operationally sound. Any *a priori* assumptions as to weaknesses in multi-ethnic, polyglot, non-citizen, militarily diverse armies must be abandoned. In this chapter, we aim to assess the evidence for unit and army cohesion in Carthaginian forces on their own terms. We begin with an analysis of how characteristics, such as peer and task cohesion, helped them accomplish their goals, and then move into a discussion of the contribution of vertical or rank cohesion.[3] The end result is a detailed examination of how Punic armies cohered and of the different social and military influences and structures which impacted on soldiers.

Horizontal and Peer Cohesion

Looking for the effects, or absence thereof, of horizontal unit cohesion in Punic armies is not an easy task. Modern military theorists have the advantage of being able to speak to soldiers and veterans, compile large datasets, and base their studies on these. For the historian of Carthage, we are left only with the imperfect reports of Greek and Roman historians, generally writing from a perspective hostile to the Carthaginians.[4] What little ancient judgement we find which touches on this topic is from Polybius, who saw the 'non-native' qualities of Punic armies as a liability. Thus, for him, the Romans could 'turn a defeat into overall victory' because 'their country and their children are always directly at stake for them' and thus they have a stronger reason for continuing the fight (Polyb. 6.52.6–7).

This reasoning, and the supposed logic behind it, has held sway in military thinking since the ancient world. As we saw above, theorists such as Machiavelli echoed the view that national troops were always the best choice. But recent exploration of the various aspects of unit cohesion in more contemporary military conflicts has shown that a simple dichotomy of quality between 'national' and 'foreign' troops is not necessarily warranted.[5] In this section, we will look for evidence of the horizontal forces which, based in modern thinking, should have been acting upon units in Punic armies, holding them together and making them effective combat units.

Because we cannot interrogate the soldiers who fought for Carthage, we are forced in this analysis to extrapolate information from what our sources tell us. This requires distillation of narrative histories and the extraction of inferred information. An *a priori* assessment leads one to see social-cohesion forces at work in units of Carthaginian citizens, Libyan conscripts and allies, as well as other ethnic units in Punic armies. The most intriguing instances of this can be found at the 'sharp end' of the military history of Carthage, that is, by looking at a number of vignettes of the battles fought by its armies.

The strength of horizontal cohesion can be seen early in the known history of Carthage. In the opening move of the Punic invasion of Sicily in 409 BC, Hannibal urged his troops forward into a dangerous city-storming situation by offering up the city for plundering. Responding to this was a group of Campanian mercenaries who wished to perform extraordinary deeds, apparently in hopes of a great reward (Diod. Sic. 13.55.5, 7). Even though this attack was repulsed, Hannibal's army later broke through the fortifications, a group of Iberian mercenaries being the first to mount the walls (Diod. Sic. 13.56.6). After taking the city, the soldiers were allowed to sack it and run rampant, plundering 'whatever of value was to be found in the dwellings', sparing only the women who fled to the temples to prevent them from setting them alight and from being looted (Diod. Sic. 13.57.2–6).[6] After capturing Selinus, Hannibal's next target was Himera. The Carthaginian army surrounded the city and broke through its fortifications, with the Iberian mercenaries again being the first to burst inside (Diod. Sic. 13.62.2). Like at Selinus, the city was given over to the troops to plunder (Diod. Sic. 13.62.4). Now, having taken Himera, which may have been Hannibal's original goal in the campaign, he dismissed his army, with the allies and the mercenaries returning to their respective homes. But before they left, the Campanians were bold enough to complain to their general that they had not received fair compensation for their services, since they were 'chiefly responsible for' the successes of the war (Diod. Sic. 13.62.5).

Within the otherwise sparse details of the campaign of 409 BC, we get a glimpse at the forces of horizontal cohesion which acted to hold units in the Carthaginian army together, even in heavy fighting. Firstly, despite the army being ethnically heterogenous, individual units seem to have been organised based on existing relationships. Thus, we see distinct units of Campanians and Iberians. Both of these performed well under pressure. By keeping these groups together in mono-ethnic units, the army benefitted from the strength of social cohesion without having to spend lengthy periods of time training and drilling. These groups of mercenaries already have a common language, approach to warfare and value system, as well as other cultural traits.[7] It is probable that the soldiers which made up these units also had a common fighting style and sets of equipment, making them recognisable on the battlefield and able to fight side by side effectively. They also benefited from peer-bonds, akin to the modern idea concept of 'the buddy principle'. That is, soldiers will fight harder and be more committed to the task at hand if they are in combat alongside someone who they know and trust. This forms what has been called primary group cohesion, a connective power that makes them an effective fighting force.[8] Creating this is not necessarily an easy task; for instance, it has been argued that the U.S. Marine Corps became a strongly cohesive military formation after desegregation only by adopting intensive measures in the boot camp to inculcate its core values ('pride, loyalty, and the shared understanding of the common purpose') that would facilitate the creation of a strong esprit de corps.[9] Having this type of cohesion already present in units is extremely helpful both for the army's performance and for its leaders, who would not have to commit as much time to building it as they otherwise would.

The ethnically distinct units in Carthaginian armies, on the other hand, already possessed coherent identities and traditions in many of the ways that formations like the U.S. Marine Corps had to develop on the training ground. For instance, we can see the presence of strong ties of common language and culture in the Campanians recruited for the Carthaginian campaigns in Sicily from 409 BC.[10] Coins struck by this group, who settled in Entella soon afterwards (Diod. Sic. 14.9; Ephorus ap. Steph. Byz. s.v. Ἔντελλα), provide evidence of their shared identity. Drachms were struck on their reverse with a helmet and the legend ΚΑ-ΜΠΑΝΩΝ, with hemidrachms bearing the legend ΕΝΤΕΛΛΑ on their obverse and ΚΑ-ΜΠΑΝΩΝ on their reverse.[11] The use of a single ethnic moniker, 'kampanwn', is indicative of an assertion of a single identity amongst the group. The use of a helmet on the reverse, one that is found in southern Italy, is also evidence of a shared martial tradition within the group; it was a symbol with which they could identify.[12] Because of this shared identity, peer bonds further strengthened these relatively small groups. Thus, at the most basic level, the combat unit, Carthaginian armies should not necessarily be seen as 'multi-ethnic' conglomerations in a pejorative way, but rather as collections of units fighting for the same goal that were internally united by traditional social and cultural bonds.

In this light, it is also worth noting the Campanians who would serve Agathocles, eventually abandoning the cause of Syracuse and setting themselves up in Messana and who became known as the Mamertines. They seized the city around 288 BC from its earlier inhabitants (Polyb. 1.7–8; Diod. Sic. 21.18). Like their cousins in Entella, they remained a cohesive group even after disengaging from their original employment, going on to exist as a political entity until the First Punic War. Again, like the earlier group of Campanians, they minted coins that identified them as a group. Both instances are evidence of the strong bonds within mercenary bands, almost certainly forged by shared experiences and language, existing social bonds, and vertical cohesion with their leaders, which is explored in more detail below.[13]

The power of task-cohesion is also in operation at the sieges of Selinus and Himera. In spite of the dangers of assaulting a city, groups of both Campanians and Iberians were willing, if not eager, to attempt it at both cities. This enthusiasm was evidently not from altruistic feelings towards their Punic commander, but rather in hopes of economic rewards. Thus, we hear that in the case of both attacks, Hannibal promised his troops that they could loot and plunder the cities. The chance of earning a reward gave the mercenaries a reason to fight as hard as possible, as a group, which is reflected in the parting complaints of the Campanians. They went to the Carthaginian leaders claiming that they had not been fairly compensated for their actions in the war, hinting that their main motivation was economic gain.

Jeremy Armstrong has recently argued, in the context of early Rome, that 'the acquisition of portable wealth through warfare would have formed a direct and tangible goal which could bring together even the most disparate entities'.[14] The profit from raids was a strong enough motivation to allow groups to cohere in high-pressure situations.[15] This had the potential to create 'brotherhoods' of raiders, for instance, in Viking society, willing to travel great distances and risk their lives in search of portable wealth.[16] By suggesting that his soldiers would receive considerable economic compensation for capturing the cities of Selinus and Himera, and offering them to his men to loot, Hannibal was creating a powerful motivating task which exerted a strongly cohesive influence on the units within his army.

The power of economic reward to maintain unit cohesion is further evident when the Carthaginians returned to Sicily in 406 BC. During the protracted siege of Agrigentum, the Punic camp ran low on food, which drove a group of Campanian mercenaries to threaten revolt. Himilco, the general in command, quelled this by offering them the valuable drinking cups of the Carthaginian soldiers and a promise to bring in new supplies to the camp (Diod. Sic. 13.88.2–3). He did this by capturing a Greek grain shipment, an act so bold that, in combination with fifteen talents worth of silver, not only satisfied his own Campanians but brought those Campanian mercenaries who, after the previous campaign, had gone into the service of the Greeks, to defect to the Punic cause (Diod. Sic. 13.88.3–5).

The strength of the forces of peer cohesion and task cohesion are also evident during several episodes at Lilybaeum during the First Punic War. In 250 BC, about 14 years after the war began, the Romans had driven the Carthaginians into the northwest corner of Sicily and began a siege of the stronghold-city of Lilybaeum (Polyb. 1.41). Both consuls for the year brought their armies against the walls, a combined force of perhaps 100,000 men. This was about ten times the number of the defenders, with the Punic garrison numbering around 10,000.[17] The besiegers circumvallated Lilybaeum, and then eagerly brought towers, catapults, and rams against the fortifications and began undermining the walls (Polyb. 1.42; Diod. Sic. 24.1.1). Himilco, the Carthaginian commander, launched regular sorties from within the city. These turned into tenacious fights, with Polybius claiming that in some of these more men were killed than 'usually are in pitched battles' (1.42). While this sounds hyperbolic, conditions within the city were dire enough that a number of high-ranking mercenary captains formed a cabal to betray their units to the Romans (Polyb. 1.43.1). A Greek mercenary, Alexon, was not swayed to participate in this plot, and alerted Himilco to it.[18] In an effort to keep his soldiers loyal, the commander offered them rewards in excess of the bounties they were already owed. This was enough that when the traitorous leaders returned to the city's walls, their own men attacked them (Polyb. 1.43).

Economic reward and the unifying task of earning it was a strong enough cohesive force to make the various mercenary units within Himilco's force stay loyal to him, and to each other, rather than to their immediate commanders. This was in the face of vast numbers of the enemy and of what must have seemed like certain defeat. This incident suggests that task-cohesion – here being economic reward – was a strong enough force to override the vertical forces of cohesion between the mercenaries and their captains (on which, see next section on rank cohesion). Indeed, the morale of the garrison was such that when reinforcements were brought into the city through the Roman blockade (Polyb. 1.44), Himilco felt confident enough to order a large sortie against the siege works.[19] Although he committed many troops, and almost emerged victorious, the savage melee ended in considerable Punic casualties (Polyb. 1.45). Yet, despite this loss, his soldiers stayed loyal, and we hear of no more talk of desertion. Finally, as the siege dragged on, a group of Greek mercenaries approached Himilco with a bold plan. On a night with a strong wind blowing into the Roman siege works, they proposed to launch another attack. Those involved in the sortie would be armed with firebrands and other implements for setting the engines alight. This effort was successful and brought to an end Roman attempts to breach the walls of Lilybaeum (Polyb. 1.48). The boldness of this plan and the mercenaries' willingness to carry it out are further proof of their cohesion.

The impact of this Roman Stalingrad on the Punic garrison are illustrative of the strength of horizontal cohesion on maintaining the fighting ability of both the army more generally, as well as the smaller units of which it was composed. Bringing in reinforcements enhanced morale by making the chance of survival seem more probable, but the cohesive strength of small units, presumably of peers, is still evident. The events of this siege support modern studies which argue for the importance of horizontal unit cohesion (especially task-cohesion).[20] In this episode, we also see the power of rewards in reconfiguring the vertical relationship between Himilco and his soldiers, as we discuss below, which offered them a clear alternative to the promises of their immediate commanders. Thus, at Lilybaeum, it is clear that the cohesive forces created by the 'task' of earning bounties from Himilco, as well as the bonds between the rank-and-file men of individual units, trumped what existed between the soldiers and their mutinous immediate superiors.

Punic citizen units were also influenced by the power of horizontal cohesion. We can say this certainly of the so-called Sacred Band. This unit was made up of wealthy and valorous citizens equipped with spectacular arms and armour, making them identifiable on the battlefield. This formation is only heard of twice: at the Battle of the Crimisus in 339 BC and the Battle of White Tunis in 310 BC.[21] Even if its battle record is not well known to us, it appears that the Sacred Band experienced strong forces of cohesion. At the Battle of the Crimisus, despite a driving rain blowing into their faces and the chaotic rout of the rest of the Carthaginian army, the Sacred Band held strong, fighting to the death as a cohesive unit (Diod. Sic. 16.80.4; Plut. *Tim.* 28.5).[22] At the Battle of White Tunis, the unit showed similar resiliency. They were stationed on the right-wing of the Punic battle-line, under the direct command of Hanno, one of the generals. They were opposed on the Greek side by Agathocles, the Syracusan tyrant, and his bodyguard of 1,000 hoplites (Diod. Sic. 20.10.6, 11.1). Once again, the Sacred Band fought a tough, losing, battle. Although their general was eventually killed in the melee, they only abandoned the field after a stout resistance, once the rest of their army had fled (Diod. Sic. 20.12.3, 7). Diodorus makes it clear that they fought harder than their compatriots.

An explanation of how and why the Sacred Band was such a cohesive unit is not immediately clear. We do not know how well they were trained, but they could have received more drilling and practice in martial skills than the typical citizen-soldier from Carthage, or perhaps even more than the allied and mercenary troops.[23] They also may have been more used to being on campaign than other units, it being noted that they had been 'drawn from the ranks of those citizens who were distinguished for *valour and reputation*, as well as for wealth' (Diod. Sic. 16.80.4; Plut. *Tim.* 28.6). Although we only hear of the Sacred Band twice, and so can only speculate on which campaigns they may have otherwise been involved in, we do know that for some period of time the Carthaginians rewarded their citizen-soldiers with armbands for every war in which they fought (Arist. *Pol.* 7.2.10 = 1324b). This implies that those who served in the army did so for more than just a single campaign. Aristotle's note was written in the fourth century BC, probably making it applicable to those who were selected for the Sacred Band. The collection of these rings, or other outward insignia, could be regarded as a task for the members of citizen units, as markers in the construction of reputation, and as tangible rewards for valour and time spent on campaign.

There may have been more to the cohesive forces which held the Sacred Band together. The power of esprit de corps, while argued as fairly weak by some modern commentators, has historically been considered very important, and cannot be discounted here.[24] By singling these soldiers out from the other citizen troops, as well as from the other elements of the army, the Carthaginian leadership created an identity around which individual troops could cohere; they created new bonds of social cohesion between the men. Indeed, so much so that this group was described as *hieròs*, sacred. But we are not given much to go on by our sources. Modern scholarship has rightly dismissed any connection with the Sacred Band of Thebes, which apparently consisted of 150 pairs of lovers.[25] As we have noted Diodorus, identifies the men who made up this unit at the Crimisus as possessing bravery and fortitude. Similarly, he calls them 'picked men' (*epilektoi andres*) at White Tunis, implying that they were especially valued in combat (20.12.3). The latter may imply that it was an ad hoc unit formed on campaign with veterans who had been noticeably effective. This is improbable, however, since Diodorus elsewhere in his work uses the term 'picked' (*epilektoi*) specifically for a variety of elite units.[26] It is also tempting, based on their description at the Crimisus by both Diodorus and Plutarch, that they could have been differentiated because of their or their family's wealth.[27] Alternatively, they may have operated as a bodyguard unit (see below), perhaps the 'sacredness' of the unit was due to oaths that they had sworn to their commander or with one another. If this was the case, it helps to explain their 'fight to the death' mentality. This was bolstered by the various cohesive forces present within the unit, both horizontal (peer, esprit de corps, ideological?) and vertical (loyalty or oaths to their commander). As citizens, these bonds may have been strengthened in light of the fourth-century pride that Carthaginians took in serving on campaigns, as seen by the awarding of armbands as we noted above. Unfortunately, due to the sparse nature of our sources, we cannot push such speculation far.

Looking beyond the evidence for horizontal cohesion in individual units, Punic monetary policy may have been a broader source of social cohesion in Carthaginian armies after monetization. The campaign of 409 BC against the Greek cities of Sicily probably resulted in the minting of the first coins by Carthage. These were tetradrachms of the Attic measure.[28] They carried the legends 'QRTḤDŠT' or 'MḤNT', meaning, respectively, 'Carthage' and 'the camp'.[29] The second of these indicates that they were minted 'only for a specific purpose', that being to pay the soldiers: mercenaries, allies, and citizens.[30] These legends, while making it obvious who minted the coins, were also a constant reminder for their bearers from whom they earned their pay. This had the possibility of creating a shared identity amongst the veterans of Carthaginian wars, identifiable by the money they carried and spent, and they would be regularly reminded of their service and rewards. A similar cohesive force may have been created by the rebel army operating in North Africa during the Truceless War. They minted coins with ΛΙΒΥΩΝ on them, with the occasional M, A, and Z. This represented 'The Libyans' and the individual letters were perhaps some of the rebel leaders.[31] It is difficult to quantify how strong these identities would have been, but there is little doubt that it would have contributed, in some way, to the cohesion within Punic armies.

By looking at a number of vignettes into the functioning of Carthaginian armies, it is easy to see that they were held together, at least in part, by a variety of horizontally cohesive forces. Social cohesion existed amongst groups of soldiers who fought alongside their countrymen, whether they were Carthaginians, Iberians, Campanians, or from elsewhere. This peer cohesion was enhanced by such actions as the minting of socially identifying coinage and the creation of special units, such as the Sacred Band. Importantly, though, this was not the only horizontal force at work. Task cohesion appears to have been as strong in Punic armies as has been argued for more modern fighting forces. This was strong enough to drive groups of mercenaries to abandon their own leaders in favour of the Carthaginian cause at Lilybaeum, and inspired others to engage in the risky business of breaching the walls of Greek cities. Understanding that these forces acted upon Punic armies, and how they acted, helps us to better understand how those armies functioned. It also makes us cast further doubt upon the judgement levelled upon the effectiveness of these fighting forces, making us eschew the notion that 'non-national' units were less effective and reliable than citizen-units.

Vertical and Rank Cohesion

The elite perspectives of our ancient commentators shape our surviving accounts of the Carthaginian military. They frequently display top-down views of the relationship between commanders and the common soldiery. Polybius, for example, tends to characterise all common soldiers (not just those in Carthaginian armies) as an irrational mass, requiring educated leaders to restrain and control their impulses and bend them to the will of command through the exercise of reasoning power.[32] He heaps praise on officers who most effectively lead their men through personal charisma and authority, combined with a scientific understanding of all matters relating to generalship, and he cites both Hamilcar Barca and his son, Hannibal, as exemplars.[33]

There are reasons to think that this view has some merit. From the moment that an ancient army was assembled, the commander began to develop a relationship with the troops. A commander could promote cohesion by fostering loyalty, promoting courage and commitment, instilling discipline and order, and ensuring the men acted as an effective fighting force. The more prosaic concerns for nourishment and pay could be essential in promoting respect for the commander and maintaining morale. Indeed, almost all that a commander did and how he acted could have an impact on the way that the troops regarded him. The Carthaginian army benefited from the fact that Punic generals were not normally time-limited, but remained for the duration of campaigns, intrinsically providing a continuity and familiarity with the troops as the war progressed.[34] The longevity of command meant that it was possible for leaders to build up relationships of respect and trust. This was particularly important in Carthaginian armies, since these were often assembled from diverse nationalities, and where individual and group motives for serving might range from pay to compulsion. The exemplary loyalty of Hannibal Barca's men in Italy (218–203 BC) was a high point in military cohesion and effectiveness. His father, Hamilcar, was also able to keep an army in the field in Sicily for long periods (247–242 BC), without sufficient pay and despite a military stalemate. On the flip-side, repeated incompetence displayed over time caused officers to become utterly worthless in the eyes of the soldiers. After serial defeats at Agrigentum (262 BC), Mylae (260 BC), and Sulcis (258 BC), the responsible commander, Hannibal, was apparently executed by his own men.[35]

Carthaginian generals attempted to reach out directly to the soldiers in various ways. The use of camp assemblies and speeches allowed the commander to address and motivate the army, encouraging a sense of common purpose and showing the soldiers that he believed in their qualities and their chances of victory. Thus, in 250 BC, Himilco roused the enthusiasm of the troops for a major sally against the Roman besiegers at Lilybaeum with a speech and promises of rewards for valour and other bonuses (Polyb. 1.45.2–3). Hannibal Barca is recorded as providing motivational speeches on numerous occasions.[36] Polybius (11.19.5), indeed, emphasises Hannibal's ability to make the men listen to his commands. These efforts by commanders to address the troops in military assemblies indicate that one essential element to effective vertical cohesion is communication, which flows down from command and up from subordinates. The impact on cohesion is most easily seen in the practicalities of transmission of orders and in encouraging men, but also in officers' attempts to understand their mood and even listen to their suggestions. This goes beyond the issue of the most basic commands, which might be conveyed by signals, trumpet blasts, and other such non-verbal mechanics, to more complex and abstract matters of extended explanation, reasoning, and discussion.[37]

The efforts of Carthaginian generals to promote themselves directly to the army by making visible attempts to address the men before operations, using pre-battle speeches, was not straightforward, however, since the polyglot armies of Carthage posed particular problems. Polybius (1.67.8) thought that it was impossible for a Carthaginian commander to speak all of the languages of his men. Though, there was an alternative stereotype of Punic commanders which appears in Plautus' *Poenulus* (prologue, 114), where Hanno the Carthaginian is described as knowing all languages (*is omnis linguas scit*).[38] But Polybius' (1.67.9) criticism that statements at army assemblies had to be repeated in four or five different languages cannot be ignored. He notes that generals often had to use local commanders and interpreters as conduits for information and the relay of orders.[39] Thus, at Zama, Hannibal's officers each made speeches to their contingents before the engagement.[40]

As we have seen, the organization of armies along ethnic divisions was a typical feature of Punic armies in all periods, and it seems likely that this at least facilitated internal communication *after* the point of translation.[41] Curiously, however, Polybius (1.67.4) argued that the Carthaginians deliberately fostered the differences between ethnically and linguistically distinct contingents as a means to make them more pliable, less liable to disrespect their officers, and less able to combine in seditious ways. In such a system, the generals and the troops would have had to rely on the abilities of translators and bi-lingual officers (where they existed) to promote vertical cohesion through such mediated channels of communication. This was potentially subject to abuse during moments of sedition, as appears to have occurred during the unrest at Sicca in 241 BC, where some officers and translators were reported to have distorted the words of the general, either due to misunderstanding or for malicious political ends (Polyb. 1.67.11). For this reason, the fracture of communication in times of crisis and mutiny could be potentially disastrous for the cohesion of the army. The general at Sicca, Hanno, found it almost impossible to calm the turbulence in the camp, despite support from subordinate officers (Polyb. 1.67.3). Indeed, such was the failure in communication on this occasion that the men became 'distrustful of their divisional officers (*kata meros hēgemones*), and highly indignant with the Carthaginians' (Polyb. 1.67.13).[42]

Nevertheless, we should be careful not to over-emphasise structural weaknesses in military communication. Polybius also suggested that some familiarity with Punic existed among more veteran troops (1.80.5–6), which allowed the Gaul Autharitos, who was reasonably fluent, to dominate the assemblies of the mutineers during the Truceless War.[43] It is possible, too, that many Libyans in the armies of Carthage had an acquaintance with Punic.[44] Indeed, aside from the events of 241 BC, it is striking how rarely occurrences of such breakdown are reported. For the most part, the Carthaginians proved adept at communicating with their subjects, allies, and employees. Indeed, there were further ways that commanders could promote the cohesion and military effectiveness of their armies.

Because of the practicalities of communication across language barriers, it was difficult for generals to maintain a personal touch in large armies of tens of thousands of men. Thus, speeches to the whole might be augmented by more individual contact. Hannibal, prior to the battle of Trebbia, put 200 men under the direct command of his brother, Mago, and at his tent praised their bravery, instructing each one to co-opt ten dependable colleagues from their own 'companies'.[45] The force was trusted with undertaking a vital ambush in the battle (Polyb. 3.71.5–9; Livy 21.54). The cascaded approach adopted by Hannibal here suggests that he understood how cohesion and military effectiveness might be promoted through the personal contact of interlinked small groups. We have already noted the presence, at various levels, of subordinate officers, who acted as the primary conduits of communication, but our understanding of the Carthaginian chain of command is limited. We lack the detailed accounts to allow us to reconstruct it with the same detail or certainty as is possible for the Roman or some Hellenistic militaries. The matter is complicated by the fact that most Carthaginian armies were ad hoc assemblages from a variety of sources of manpower: mercenaries, allies, subjects and citizens of diverse ethnic origins and military traditions. It is possible that there never was a typical Carthaginian army, but that there were substantial variations over time and, even, between armies in the same conflict.[46]

Nevertheless, there is suggestive evidence of the roles performed by the intermediate tiers of command in promoting the cohesion both of the army and its constituent parts. By drawing on modern military studies that explore the concept of 'leader-subordinate', 'vertical', or 'rank' cohesion, we can begin to evaluate the hierarchical relationship between the soldiers, their immediate officers, and the higher organs of command.[47]

An 'officer' can be said to be a member of two hierarchical groups: a primary assemblage of subordinates that he himself leads, such as those in a unit or detachment, and with whom he has most personal contact, and a somewhat larger secondary group, such as a division of the army.[48] The officer is usually part of a group of fellow officers who collectively command that larger, secondary, portion of the army. They, too, are all subordinate to a commander, and constitute his primary group. The superior officer may himself belong to a higher group under a more senior commander, and so on. At each level an officer is often responsible for the effectiveness of the group under his command, which includes its ability to discharge assigned tasks, its sense of identity, and its commitment to the army and mission objectives. Wherever he sits in the chain of command, the officer acts as a 'linking-pin' between the primary group of immediate subordinates, the secondary group represented by fellow officers of the same rank, and a superior commander.[49] The linking-pin theory helps to describe how, at each hierarchical layer of an organisation, the leader and immediate subordinates overlap and combine effectively, and how vertical cohesion might be enhanced or undermined by the relationship between individuals and groups. If the individual officers are unable to knit the constituent parts of their own groups together then an army is likely to be less effective on campaign.

Although we have seen Carthaginian generals attempting to connect to the common soldiers through camp assemblies and individual interactions, for the most part, the general's primary group was not the common soldiery, but the senior officers. Polybius argued that a commander should consult with those who he can trust and who are intimately involved in implementing the plan of operations (9.13.2–3). In the Carthaginian army, this role was provided by an upper echelon of officers who would have been tasked with commanding particular divisions in battle.[50] Some might have more specific responsibilities, like 'officer of supplies' (*ho epi ton leitourgion tetagmenos*, Polyb. 3.94.4). In Hannibal's army this person not only arranged the commissariat but also oversaw foraging with a covering force (Polyb. 3.102.1, 5). This role appears to have come with considerable authority and responsibility for mounted operations, with the same officer commanding the heavy cavalry at Cannae (Polyb. 3.114.7, 116.6–8; Livy 22.46, 48).[51]

Such officers did not just passively implement commands, but also attended the commander's council, the *synedron*. In 250 BC, officers (*proestōtes*) were called to a military council in Lilybaeum to share in Himilco's deliberations on how to forestall treachery and desertion among some elements of the garrison (Polyb. 1.43.3, and see above). Polybius presents us with several examples of Hannibal's *synedron* in action, most noticeably in planning the crossing of the Alps (Polyb. 9.24).[52] On another occasion, before Trebbia in 218 BC, *synedroi* discussed Hannibal's plan of engagement (3.71.5); similarly, after Trasimene in 217 BC, a gathering of senior officers planned the army's next moves (3.85.6). Such councils were opportunities for the officers to offer criticism and advice, perhaps the most famous being the urgings of Maharbal to Hannibal after Cannae to march on Rome, and his response, when his general showed little enthusiasm, that Hannibal 'knew how to win a victory, but not how to use it' (Livy 22.51).[53] Likert called such internal discourse and debate a 'constructive use of conflict', arguing that where organisations exhibit a positive self-criticism capacity, they seem to be more resilient.[54] Indeed, such discourse enhances loyalty and coherence.[55] The importance of the senior officers in the conduct of operations and in the effectiveness of the army is clear from the fact that in 218 BC the Romans had demanded that Hannibal and his *synedroi* should be handed over after the attack on Saguntum (Polyb. 3.20.8).

So, the council was the primary group for the commander in chief, comprising the officers with whom he was most regularly and intimately connected and whom he trusted to execute the agreed plans. Indeed, some of these men were so close as to be termed 'friends', *philoi*.[56] The deliberative and supportive nature of the *synedron* thus provided a different role than the army assembly, which was predominantly a forum for the commander to explain, in broad terms, the aims of the coming operation and to encourage the men. The distinction is clear at Lilybaeum, where, after testing the morale of the army with an inspiring speech, Himilco called a council in which he discussed the plan to sally against the Roman siege-works, assigned each officer his role and position in the assault, determined the watchword and the hour. He then gave the order to the commanders to ready their contingents by the morning watch (Polyb. 1.45.5).

The maintenance of an effective chain of command below the general's immediate circle of friends and council members was a more complex business. This was due to the multi-ethnic and polyglot nature of armies that the Republic deployed. While divisional or other senior officers were usually Carthaginians, ethnically distinct contingents tended to be commanded by native leaders.[57] Where forces were raised from specific allies or subject communities, or hired as bands, it is likely that these often arrived at the camp already partially or fully pre-bonded to their own leaders.[58] The Carthaginians regularly, it seems, took the easiest course and relied on such officers to act as the 'linking-pins' between the high command and their primary groups (the contingents). This, however, provided several challenges, principally in terms of integration, language (as we have discussed), and elision of war-aims, but it is striking that some non-Carthaginian officers were given considerable responsibility. Some could act as leaders of particular ethnic divisions or even entire arms (such as a wing in battle or the whole cavalry), and thus were likely to have attended the *synedron*.[59] In part, this was a recognition of their importance within their own ethnic context – as communal leaders and 'warlords' – which it was politic for the Carthaginians to acknowledge and to exploit, but it was also sometimes a selection based on the military qualities of the leaders themselves, and thus, partially, meritocratic.[60] It was the case, for example, that the talented Numidian prince Massinissa was, at one point, given the command of 3,000 of the 'best cavalry' to harass the Roman armies in Spain (Livy 27.20).[61] He had also played a key role in the campaign that resulted in the defeat of two Roman armies and the deaths of their commanders (Livy 25.34–5).

When our ancient narratives inform of us of the names (and ethnicities) of officers, we tend to find them commanding contingents numbering no fewer than 500 troops, but it is worth noting that below such leaders there were likely to have been junior officers who commanded smaller elements.[62] We have some references to cavalry forces of around 500, sometimes further subdivided into squadrons, probably of 150–200 men, that may have been further divided into groups of around 30.[63] Infantry contingents were also apparently subdivided, large contingents such as 2,000 Gauls at Tarentum might be split up for operational reasons – in this case into three groups, commanded by Celtic leaders (Polyb. 8.30.1) – but we also hear of contingents of around 500 men.[64] Both Livy and Polybius also mention forces of roughly maniple size (c. 150 men) in the third century BC, suggesting further or alternative articulations of ethnic contingents.[65] All of these subdivisions would have been commanded by junior officers. It is unclear how regular or formal such units were or how the leaders were selected, but we do know of a few Punic military terms for the leaders of such contingents: *rab met*, 'commander of a hundred', *rab halus* 'senior equipped soldier', and possibly *hanno rab selos* or *hanno rab*, meaning 'troop leader' or 'chief shield bearer'.[66] It is possible that these Punic terms rarely penetrated the veil of language used within contingents, or at best sat alongside the troops' own native terminology. Nevertheless, it was such junior officers who were probably relied upon to command the groups of 200 or 300 men scattered in ambush across the battlefield near Gerunium in 217 BC.[67]

The various tiers of the army, from generals to small unit leaders, all the way down to common soldiers, were connected by a series of personal relationships that formed through familiarity and the daily necessities of command. The vertical cohesion that this engendered was at each step direct and relatively intimate. Consequently, the willingness of the subordinates to trust and follow their commander was often based on two key aspects: a perception of the superior's soldierly virtues, competence, and conduct (the *instrumental aspect*) and of their officers' behaviour towards their men (the *affective*).[68]

The instrumental aspects that enable a leader to develop a good relationship with his troops, in part, are based on their expectations of what an officer should be able to do. Simply put, if an officer knows the business of war, the correct military terms and orders, the best methods to engage the enemy or conduct an operation, how to organise the men and keep them supplied, then soldiers will be more positive in following him. Thus, despite initial scepticism among the ranks, when Xanthippus drew up the divisions, manoeuvred them, and issued orders 'in the orthodox military way', the troops' confidence became much enhanced, and they cheered in approval (Polyb. 1.32.7). On his arrival in Sicily in 212 BC, the Liby-Phoenician officer Muttines was described as an 'active soldier, thoroughly trained in the science of war under Hannibal' (Livy 25.40), who soon demonstrated his abilities and inspired the men he led. An officer's reputation could be enhanced by displaying a range of additional qualities such as bravery, good judgment, luck, divine favour, persuasion, foresight, energy, and comradeship. Himilco Phameas, for example, could be praised for possessing 'great personal vigour, and, what is most important in a soldier, being a good and bold rider' who was 'by no means timid' (Polyb. 38.6.1, 3). According to Livy (21.4), young Hannibal Barca's demeanour in the field when he was still a subordinate officer to Hasdrubal was noteworthy: 'he was fearless in exposing himself to danger and perfectly self-possessed in its presence. No amount of exertion could cause him either bodily or mental fatigue; he was equally indifferent to heat and cold; his eating and drinking were measured by the needs of nature, not by appetite; his hours of sleep were not determined by day or night, whatever time was not taken up with active duties was given to sleep and rest, but that rest was not wooed on a soft couch or in silence, men often saw him lying on the ground amongst the sentinels and outposts, wrapped in his military cloak. His dress was in no way superior to that of his comrades; what did make him conspicuous were his arms and horses. He was by far the foremost both of the cavalry and the infantry, the first to enter the fight and the last to leave the field'. These qualities appear to have made Hannibal the perfect link-pin officer: 'never was there a young character more capable of the two tasks so opposed to each other of commanding and obeying; you could not easily make out whether the army or its general were more attached to him. Whenever courage and resolution were needed, Hasdrubal never cared to entrust the command to anyone else; and there was no leader in whom the soldiers placed more confidence or under whom they showed more daring' (Livy 21.4).

Vertical cohesion, therefore, was promoted by the admiration of the men for their officers and could be enhanced through their shared experiences, personal interactions, and the past performance of the tasks assigned them.[69] It appears that Hannibal's officer of supplies in 217–216 BC, through regular supervision and protection of foragers at Gerunium (Polyb. 3.102.1, 5), and his involvement in executing some of Hannibal's bold stratagems, such as the famous distraction of Fabius' army with flaming oxen (Polyb. 3.93.4), created sufficient bonds of shared experience and trust for the cavalry he commanded at Cannae to display exemplary cohesion. In the aftermath of its rout of the Roman right, this force was able to perform the difficult task of breaking off pursuit in order to execute decisive attacks on the rear of the Roman left and then the centre (Polyb. 3.114.7, 116.6–8; Livy 22.46, 48). Shared experiences and past relationships with soldiers were valuable instruments of cohesion at times of dissent. In 250 BC, at Lilybaeum, after the tip-off from Alexon, a certain Hannibal was chosen by his commander to help negotiate with a potentially disaffected contingent of Gauls because he had commanded them before (Polyb. 1.43.4). Conversely, the removal of trusted officers could undermine the commitment of troops. In 210 BC, the general Hanno tried to diminish the growing influence of the militarily successful Muttines by replacing him with his own son. This created considerable animosity, however, and caused a Numidian contingent based at Agrigentum to join Muttines in betraying the city to the Romans (Livy 26.40).[70]

The close bonds between commander and soldiers might be promoted by their role in combat. At White Tunis, the Sacred Band fought directly under the command of one of their generals. They appear almost like a bodyguard unit in this battle, as they were said to have fought alongside Hanno until he was killed, and that others led the unit afterwards (Diod. Sic. 20.12.7). They may have functioned in a similar manner at the Crimisus. There, the entire unit was wiped out, having formed the vanguard of the army crossing the river. Although we do not hear of them dying, explicitly, neither of the generals are heard about after the battle, and the Carthaginian leadership was forced to send a man named Gisco to Sicily as a new commander. It is thus possible that the Sacred Band was killed while acting as a bodyguard unit for their commanders at the Crimisus. If this was, indeed, the role of this unit, the task of protecting their commander(s) could have exerted an exceedingly strong cohesive force on them. This helps to explain what drove them to fight to the death at the Crimisus and withdraw from the field at White Tunis only after the rest of the army fled. Forces of vertical cohesion would also have been exerted here, perhaps even stronger than within the rest of the army as the commanders were present. Even if the Sacred Band were not an actual bodyguard, the action of the general, Hanno, at White Tunis suggests that a personal display of bravery would capitalise on and promote the cohesion of this force. For, as Diodorus (20.12.3) asserts,

> Hanno, who had fighting under him the Sacred Band of selected men and was intent upon gaining the victory by himself, pressed heavily upon the Greeks and slew many of them. Even when all kinds of missiles were hurled against him, he would not yield but pushed on though suffering many wounds until he died from exhaustion.

The officers and soldiers of the Sacred Band subsequently displayed similar fortitude, 'stepping over the bodies of their own men as they fell, and withstood every danger' (Diod. Sic. 20.12.7), at least until the battle was seen to be completely lost.

Vertical cohesion is also influenced by the affective qualities of officers, for however charismatic and soldierly a leader may be, it is equally important that they are able to ensure the comfort and well-being of the men, as far as it is in their power to do so. Leaders often earn the respect and devotion of the men under their command, and are judged to be 'good officers' through their management of external factors. Pressures on the men might come in the form of higher command directives, logistical or environmental pressures, or other events. The more successful officers appear to address such issues through various strategies of incentivisation, preparation, intercession, and amelioration. Unfortunately, it is rare for the ancient sources to discuss such solutions, at least at the junior-officer level. The narratives tend to focus on the commander-in-chief, who often gains the credit for providing incentives and solutions to the army's problems. This is inevitable, especially given that rewards and bonuses to the men appear to have flowed from the general in most situations.[71] According to Polybius, in the winter of 220–219 BC, Hannibal, 'by the generosity he now displayed to the troops under his command, paying them in part and promising further payment, inspired in them great good-will to himself and high hopes of the future' (Polyb. 3.13.8; cf. Livy 21.5). This was also key to sustaining the task-cohesion of the army in Sicily during the difficult years of constant exertion in defence of Lilybaeum, as we have already seen, and during the stalemated warfare in the west of the island (250–242 BC). We hear of bonuses being offered by the commanders for loyalty in 250 BC and further promises 'in critical situations' also appear to have been made to the men (Polyb. 1.66.12; cf. 1.43.4–5, 45.3, 67.12).

In addition to offering rewards, demonstrating due care for the troops' health and well-being was a significant way to promote bonds of trust and loyalty. Thus, Polybius explains how, in 241 BC, a Carthaginian officer was chosen by the veterans at Tunis to be a mediator on the issue of their outstanding pay: '[they] being very favourably inclined to Gesco, who had been a commander in Sicily and had been full of attention to them in other matters and in that of their transport [from Sicily to Carthage]' (Polyb. 1.68.13). Hannibal Barca is recorded as taking great care over the troops' recovery from exertions and battles, such as after the Alps, Trasimene, and Cannae, even though he might ruthlessly push them through adverse conditions (Alps, Anio).[72] It is a measure of his success at garnering the loyalty of his men that such exertions do not appear to have substantially affected their morale and fighting capacity.[73]

Studies of modern armies have noted that officers who approach superiors to win concessions, privileges, and shares of supplies or rewards, gain stronger influence with men and enhance the sense of the group.[74] It is this affective aspect for which we have some limited evidence of the role of junior officers in the Carthaginian military. Those who, through intercession, changed command policy might be rapturously endorsed by their men. Our best example is the story of the Lacedaemonian soldier Xanthippus, who arrived in Africa during the failing campaign against Regulus and, whose criticisms of command decisions spread beyond his friends and peers, to reach the ears of the commanders. His solutions were presented formally to the commanders and then to the assembly of the army and, having demonstrated his instrumental abilities (by knowledge of command and manoeuvres), and thereby overcoming sceptics in the army, the men became eager to be led by him in the way he suggested. The resulting affective changes in the method of campaigning gave them confidence.[75] Indeed, at the approach of the Roman army, they gathered in groups to call out his name, asking him to lead them into battle (Polyb. 1.33.4).

The Xanthippus episode illustrates a more general and significant openness of Carthaginian commanders to the advice of subordinates. It is a feature of the Carthaginian system that appears several times during the wars with Rome, where interventions and advice by subordinates had a positive impact on the course of campaigns or on the discipline, loyalty, and cohesion of the army.[76] At Lilybaeum, it was a Greek soldier, Alexon, who first approached the general with the warning of potential treachery (Polyb. 1.43.2), and later some Greek mercenary officers advised Himilco to set fire to the Roman siege-works during a gale (Polyb. 1.48.3). In 249 BC, Carthalo listened to the warnings of his steersmen about an impending storm, preserving his fleet while condemning two Roman squadrons to its destructive force (Polyb. 1.55.5–8).[77] As mentioned above, Maharbal famously offered advice to Hannibal to march on Rome after Cannae, but it seems that Hannibal preferred to listen to his other officers who urged him to rest their tired men after the exertions of the battle (Livy 22.51.1–4; cf. Plut. *Fab. Max.* 17.2; Cato *Orig.* 4.13).[78] As we have seen, Polybius presents several examples of campaign planning where subordinates' views are advanced and discussed in the *synedron* of Hannibal, the one which took place prior to the march across the Alps appears as a model of open debate and policy consideration (Polyb. 9.24; cf. 3.71.5).[79]

However, the interests of the men and their immediate officers might not always align with those of the high command. Intercession could be a powerful tool for the disaffected. In the affair at Lilybaeum, Polybius reveals how in return for the garrison officers' planned betrayal of the city to the Romans, their deal apparently included promises of rewards to their men (Polyb. 1.43.3). Those officers at Lilybaeum who entered the Roman camp expected their men to desert when they called to them from outside the walls, but instead the men replied with missiles (Polyb. 1.43.1, 6). On this occasion, the deserters' conspiracy had been forestalled by Alexon and other loyal officers who had been able to extract from the Carthaginian commander an offer of bonuses for themselves and their men, which allowed for the reconfiguration of vertical loyalties to align with task-based inducement of continued service for economic gain. Nevertheless, mass desertion, where officers and their men collectively abandoned the army, arguably was the most emphatic break of the link-pin system.[80] In such cases the bond between officers and their primary group appears demonstrably stronger than that of the officers to the secondary group of the Carthaginian high command and the state. Appian (*Pun.* 108) gives a rare example of the deserters' rhetoric of intercession, when Phameas, at the point of his defection to the Romans during the Third Punic War, makes a speech to his subordinate officers (*iliarchoi*). He claimed that 'I have made terms for myself and for as many of you as I can persuade to join me. You have now the opportunity to consider what is for your advantage'. Evidently, the promise to secure terms for his followers was meant to influence their choice. Phameas was a Carthaginian (App. *Pun.* 97), but in the majority of cases of recorded desertions, the leader and his men were non-citizens, having ethnic identities that were not 'Punic', as well as immediate and communal loyalties. Such factors may have allowed them to more easily pursue their own interests than those of Carthage.[81]

It was essential, therefore, that the Carthaginians fostered links at this vulnerable point in the chain of command. We have already encountered instances of financial rewards promised to both men and officers during periods of crisis, but honour and status were also valuable commodities. Cassius Dio (13.54.7) asserted that Hannibal

> showed excessive honour to any of whom he stood in need; for he considered that most men are slaves to such distinction, and he saw that they were willing to encounter danger for the sake of it, even contrary to their own interests.

Indeed, his father, Hamilcar Barca, had assured the loyalty of the Numidian chieftain Naravas, and his 2,000 followers, with the betrothal of his own daughter (Polyb. 1.78.1–10). When Bithyas deserted to Carthage during the Third Punic War, he was quickly rewarded with the command of the Cavalry of the Republic (App. *Pun.* 111, 114, 120).[82]

Intercession to shield soldiers from arbitrary decisions, or at least to see fair play in terms of expectations and the terms of service, had to be balanced with the need to maintain discipline and obedience to orders. Since the eighteenth century, soldiers have often been subject to regimentation, uniformization, and strict control over their actions. These measures were thought to create cohesion. However, ancient armies, especially those constructed of disparate forces from different cultures and ethnicities, were significantly less exposed to such pressures and conditions. Military discipline was difficult to enforce in such armies. On the one hand there were soldiers who regarded themselves as fellow citizens, with civic rights and guarantees, while on the other, there were externally recruited warriors and mercenaries, who usually had only a loose allegiance to the employing state. The discipline of the Carthaginian army traditionally had been instituted under Mago (Justin 19.1), but its ordinances are virtually unknown.[83] Consequently, vertical cohesion required a degree of negotiation and collusion between commanders and the common soldiers. When 300 Numidians withdrew from the army at Himera in 212 BC, their commander, Muttines, went 'to reason with them and recall them' (Livy 25.40). In the process of give and take, leading by example was sometimes required. An illustration of this, though possibly a trope, was the action of Xanthippus during the Battle of Bagradas in 255 BC, who rode along the lines turning back any infantrymen who were fleeing. The story goes that when one soldier told him that it was easy for a horseman to urge others to danger, Xanthippus dismounted and continued to exhort the men on foot.[84] Nevertheless, it appears that officers were required to impose harsh discipline at times of operational necessity, thus during the surprise attack on Tarentum, Hannibal instructed his officers 'to keep their men in close order on the march and to punish severely all who left the ranks no matter on what pretext' (Polyb. 8.26.9; cf. Livy 25.9).

The attack on Tarentum in 212 BC provides us with several indications of the roles of officers in ensuring the cohesion of the force and the execution of Hannibal's will. Firstly, as we have seen, outside of the town, Hannibal gave strict orders to curtail the initiative of the troops (and officers), to punish any who left the ranks, and to follow his direct orders (Polyb. 8.26.7–9). Once inside, Celtic and Carthaginian officers led various contingents to seize particular objectives and kill any Roman in their path (Polyb. 8.30.4). Finally, certain officers were given the task of pillaging houses that were not explicitly identified as owned by Tarentine citizens, while the main body was held in reserve (Polyb. 8.31.6). Cohesion was essential in the potentially chaotic circumstances of a night attack and subsequent sack, and Polybius' narrative emphasises the discipline and obedience that Hannibal attempted to instil in the army through his officers.

Modern Western military chains of command are often highly stratified and articulated in ways that ancient armies rarely approached. Consequently, when attempting to identify the nature and extent of vertical cohesion in the Carthaginian army, we need to mindful of this difference in describing its command structure. Regardless, the link-pin theory has allowed us to think about vertical cohesion as a mediation between tiers of command and a structuring principle in the relationships between officers and men. Vertical cohesion was strengthened by effective communication and by the officers' competence and care for the men. This was essential in a multi-ethnic force, where collusion rather than discipline was the most significant method of exerting control.

Conclusions

Through this chapter, we have shown that there were considerable forces that acted on Carthaginian armies to keep them together, functioning and cohesive. Vertical and rank cohesion was exerted through the actions of leaders, whether generals or more junior officers, by gaining the trust of their men and inspiring them with their own actions. Although the evidence is scant for the structure of Punic armies below the high command, we have argued that there were a number of levels of officers, all of whom had a responsibility for maintaining the solidarity of their commands. Likewise, horizontal cohesive forces, such as peer and task cohesion, operated within the smaller units of Carthaginian armies. Thus, we find mercenary contingents that were willing to turn their backs on their own leaders to gain monetary reward, and, importantly, they stayed together after making that decision.

We have presented an image of Carthaginian armies that were just that, armies. By looking at the forces of unit cohesion, rather than moralising polemic and other judgments of the consistency of these bodies, it is obvious that they functioned and held together effectively. Cohesive powers within small units were based in existing social bonds (shared language, culture, identity, and familiarity), reinforced in-theatre by the need to complete the task at hand. These horizontal elements were complemented by developed vertical mechanisms, such as a chain of command and effective, inspiring, leadership. Commanders, such as Hannibal Barca, instilled confidence in their men by speaking to them directly and interacting with them in person. They were also able to delegate important tasks to 'officers' lower down the chain of command who similarly held their men's respect.

To some, though hopefully few, this image of Punic armies may be surprising. After all, voices as influential as Polybius and Machiavelli found them to be the ideal example of the weakness of non-national armies. But these types of views were based in misplaced, and possibly ethnocentric, nationalistic ideals. We have shown why these must be discarded and why Carthaginian armies must be seen as effective and cohesive.

Notes

1 This should be taken with a 'grain of salt' of course, and as Dexter Hoyos (2010): 154, has pointed out, this statement probably relates 'at most to the 3rd century'.
2 Machiavelli *The Prince* 12.76–77.
3 For definitions of these terms, see the introduction to this volume and, in general, MacCoun and Hix (2010).
4 A good overview of the methodological difficulties facing any historian hoping to write a history of Carthage can be found in the introduction to Miles (2010): 6–23.
5 On this, see the introduction to this volume.
6 The supposed nefariousness of the Carthaginians, especially their wanton defilement of temples, which showed how they 'surpass all other men in cruelty', should be read with a degree of caution. Diodorus was surely following pro-Hellenic sources that portrayed the Carthaginians in a very negative light. The accusations that they plundered temples unlike other peoples is certainly hyperbole, and probably hypocritical from a Greek point of view, cf. Hall (2018).
7 See for instance Ameling (1993), Rawlings (1996), Fariselli (2002).
8 E.g., Siebold (2007).
9 Cox (1996): 25, full discussion 24–9.
10 On the nature of Campanian (and Italian) mercenary groups, see Tagliamonte (1994).
11 Lee (2000): 4–8.
12 Tagliamonte (1994): 138, identifies them as of a type in use in Campania and Lucania in the early fourth century.
13 A nuanced discussion of both Campanian Mamertine and Entellan identity and language can be found in Clackson (2012). We note the important point that Campanian mercenaries may not have originated from the same communities or even spoken the same dialects of Oscan, but nevertheless we suggest that such differences may have been relatively minor hindrances to unit cohesion, at least compared to the greater cultural and linguistic gulfs between Campanians and other broad ethnic groupings such as Gauls, Iberians, or Libyans, present in the Carthaginian army. Identification with these broader ethnic categories appears to be at play in the self-identification of Mamertini or Kampanwn in the coinage issued by these mercenary communities in Sicily. On the further complexity of fusion with local population after seizure of the cities see Zambon (2008): 45–54, and Isayev (2017): 168–9.
14 Armstrong (2016): 117.
15 For instance, it was perhaps the force which held together the group who followed Dionysius of Phocaea after the destruction of their homeland by the Persians. Economic motivations led them to successful careers as seaborne raiders off the coast of Phoenicia and later in the waters around Sicily, Hdt. 6.11–17.

16 Gat (2006): 217–21.
17 The number of soldiers on both sides is unclear from our sources; see the discussion in Lazenby (1996): 123–6.
18 This was apparently not the first time Alexon had done something like this, and the preservation of his story may be down to Polybius' desire to glorify a fellow Achaean Greek, Walbank (1957): 108. This does not, however, mean that the story itself is a fabrication.
19 Walbank (1957): 108–9; Lazenby (1996): 126.
20 MacCoun and Hix (2010): 142–3.
21 Diod. Sic. 16.80.4 and 20.10.6, respectively. This could have been a short-lived unit formed in the wake of political changes which are generally thought to have happened at Carthage in the fourth century BC, cf. Hall (Forthcoming).
22 There is a discrepancy of details between Diodorus and Plutarch, including the numbers of the Sacred Band (2,500 compared to 3,000), as well as the latter not calling the unit the ἱερὸς λόχος, but as Ameling (1993): 163, has pointed out, it is 'ganz klar' that it is to the Sacred Band which he is referring.
23 Parallels are suggested by several Greek poleis who maintained and trained elite forces of citizens (termed *logades* or *epilektoi*) at public expense. See Pritchett (1974): 222–4; Tritle (1989): 56.
24 On the reevaluation of group pride, and social cohesion generally, see MacCoun and Hix (2010): 142–3; for a study which looked at this in a more traditional light, see Cox (1996).
25 Ameling (1993): 155–64. Though, see Brizzi (1995): 307.
26 Rzepka (2009): 18 n. 37.
27 Whittaker (1978): 87, refers to them as the 'political elite'.
28 Frey-Kupper (2014): 80–1.
29 Jenkins (1971–1978) described these as Carthage Series I. Lee (2000): 39–43, on the meaning of the legends. Like Lee, we reject Mildenberg's (1989): 6–8, assertion that these were indicative of a specific Carthaginian military institution. Later Punic coinage would be used as a means of asserting regional control on Sicily, cf. Prag (2010).
30 E.g., Miles (2010): 124.
31 Hoyos (2007): 139–43, with bibliography.
32 Eckstein (1995): 118–60; Champion (2004): 70, esp. 255–9.
33 E.g., Polyb. 1.75–6, 84.5–8; 9.9, 22; 11.19. On Hamilcar – Eckstein (1995): 34, 169, 174–7, 255 n. 62; Daly (2002): 117–9. Cf. Diod. Sic. 23.15.10–11 who praises Xanthippus' intelligence and practical experience of command, noting that 'armies respond to the intelligent control of their leaders'.
34 Gsell (1928): 421–2; Goldsworthy (2000): 35.
35 Polyb. 1.24.6; Livy *Per*. 17. Polybius suggests that it was the leading men or the citizens in the army who disposed of him. However, Orosius 4.8.4 suggests that he was stoned, perhaps indicative of a mutiny (cf. the lynching of officers in 241 BC, where Polybius 1.69.11 observes the only word the polyglot army came to share in common was *ballein*, 'stone him'). An alternative tradition claims that Hannibal evaded a prosecution by the authorities through a ruse, Diod. Sic. 23.10.1; Dio. Cass. 43.16–17; Zon. 8.11.
36 Polyb. 3.34, 44.9, 53, 62, 71, 111.
37 On cohesion and command signals in the Roman army, see Anders in this volume.
38 Perhaps Hannibal Barca's ability to communicate with his men rested on linguistic capabilities – he grew up in Iberia and had a Spanish wife, so may have been able to communicate directly with his Iberians, and he was certainly educated in Greek (apparently with a poor accent/style), and spoke his native Phoenician.

39 Polyb. 1.43.3, 67.1, 70.2. Cf. Hoyos (2007): 42 n. 4. Although, on some occasions, generals may have addressed the men directly. For instance, Hannibal exhorts Numidians himself and offers rewards to 'those who distinguish themselves' in the coming battle of Trebbia, Polyb. 3.71.10. When, in 218 BC, the Gallic chieftain Magalus made a speech to the whole of Hannibal's army at the Rhône, it was translated by an interpreter, Polyb. 3.44.5.

40 MacDonald (2015): 215, suggests that this was unusual for Hannibal, who is regularly depicted addressing all of his men together. Perhaps Polybius was employing a literary device to suggest the lack of coherence of the polyglot forces assembled here. This would support the observation in Goldsworthy (2000): 303, that the deployment of three lines at Zama was due to the difficulties of combining three unfamiliar armies – Mago's recently recalled from Liguria, the Carthaginian city levy, and Hannibal's army from Italy. The other possibility is that this was the norm, appearing here because Polybius had richer eyewitness sources to consult about this battle, particularly Laelius and Massinissa, amongst other survivors. It is striking that our narrators note that, despite Hannibal's long sojourn in Italy and both his (Nep. *Hann.* 13.3) and Scipio's (Livy 29.19) fluency in Greek, when the two met before the battle, they were accompanied by interpreters, Polyb. 15.6.3; Livy 30.30. This may be a mutual refusal to concede status (by making speeches in their native tongues) rather than any lack of facility in a common language; cf. Gruen (1992): 237; MacDonald (2015): 214, 292 n. 73.

41 See Pfeilschifter (2007): 31, for a similar picture of polyglot allied contingents of the mid-Republican Roman army, though Rosenstein (2012): 91–2, suggests that allies did receive some rudimentary Latin training.

42 Hoyos (2007): 23, 48–9 and n. 15, identifies these *hēgemones* as mostly indigenous officers and *kata meros hēgemones* as leaders of the various ethnic contingents.

43 Hoyos (2007): 43–5; cf. Loreto (1995): 9, 12–13, 62–5.

44 Hoyos (2007): 42, 44–5.

45 Though a larger division of the army may possibly be meant, to translate *taxis* (Polyb. 3.71.8) as 'company' here seems natural in the context of the individual nature of the selection process, and to infer that Polybius is referring to each soldier's 'primary group'. Livy 21.54 is more explicit: the others are to be drawn from cavalry and infantry units: *turmis manipulisque*.

46 A good account of the command system during the Second Punic War can be found in Daly (2002): 123–8. Variation is noted by Goldsworthy (2000): 35. Fundamental study of the changing roles and responsibilities of Carthaginian generals is provided by Wollner (1987).

47 E.g., Etzioni (1975); Henderson (1985): 107–15; Cowdrey (1995).

48 Salo (2011): 138.

49 Likert (1961): 113, developed the concept of linking-pins to describe business organisations, but his ideas are frequently applied to military contexts – see, for example, Salo (2011) and Mälkki (2014).

50 Gsell (1928): 391; Goldsworthy (2000): 35, 211–14; Daly (2002): 127–8, 133.

51 Rawlings (2016): 218.

52 Rawlings (2007): 11.

53 Hoyos (2000).

54 Likert (1961): 117–24.

55 On occasion, where secrecy was essential, such as in the night attack on Tarentum, Hannibal did not reveal or debate his plans with officers, but bade them to attend strictly to orders and do nothing on their own initiative, Polyb. 8.26.7–9. Such explicit instructions might suggest that this was not normal for his officers. Contrast Livy 25.9, who refers instead to a general assembly of the men, who are told not to act on their own initiative, but to only follow direct orders from their officers.

56 E.g., Polyb. 3.85.6. There may be some separation between a more intimate circle of Hannibal's *philoi* and a larger group of *synedroi* (Daly (2002): 127–8), but it is not clear if this was a particularly formal distinction in command terms.

57 Greek sources tend to use terms such as *hēgemones* or *proestōtes*, while in Latin *duces* and *praefecti* are regularly employed; Gsell (1928): 391–3.

58 Rawlings (2018): 157–9; cf. Trundle (2004): 105–6, for the existence of pre-bonded Greek mercenary bands for hire.

59 Rawlings (2018): 159.

60 For meritocracy in the Carthaginian military system, see Cic. *Verr.* 2.5.31, MacDonald (2015): 85.

61 Further examples in Rawlings (2018): 159.

62 It is unclear whether Xanthippus or Alexon were commanders. Xanthippus is called *stratēgos* in Diodorus 23.15.5, but Polybius' narrative suggests that he did not hold that rank. Alexon had some 'popularity and credit' with the mercenaries at Lilybaeum, suggesting that he was an influential figure in the army, Polyb. 1.43.4. At the threshold of notoriety we have commanders such as Mesotullus (App. *Pun.* 33) or Clinon (Diod. 20.38.6; 39.5), with contingents each numbering around 1,000, Bithyas with 800 (App. *Pun.* 111) and Hanno with 500 horse (Livy29.28–9, 34). The size of the Gaetulian force commanded by Isalcas in 216 BC is not specified, but it was evidently significant enough to be charged with assaulting the gates of Casilinum. It was driven off by two cohorts, so may have been of equivalent size, Livy 23.18.1; Gsell (1928): 392, doubted the episode.

63 These unit sizes are based on the terms used by our Greek and Latin authors applied to Carthaginian contingents (*ilai*, *turmae*). Livy (29.28–9, 34, cf. 25.17.3; 27.26.8) preserves an '*ala*' of 500 Carthaginian cavalry commanded by a praefectus, Hanno, who was a young nobleman. Cavalry contingents of 500 can be found in Polybius 3.44.3, though this was a specially constituted force picked by Hannibal, which Livy 21.29 described as an *ala*. Elsewhere, Livy equates an *ala* to 500 men (21.45.2; 29.28). Cf. Gsell (1928): 391.

64 Livy 21.11.8 mentions a force of 500 Libyans at Saguntum, and a Spanish 'cohort' in 211 (26.5–6); Gsell (1928): 391.

65 Polyb. 3.114.4 *speirai*; Livy 21.55, *manipuli*; Gsell (1928): 391; Daly (2002): 97, 103.

66 Definition of *Rab met*: Tomback (1978): 298 s.v. *RB* (1.3d), cf. 164 s.v. M'T; Krahmalkov (2000): 439; Donner and Röllig (1968): *Kanaanäische und aramäische Inscriften* (=*KAI*): 101.3. *Rab halus: Corpus Inscriptionum Semiticarum* (=*CIS*) 1.4823.1–2 'Adonibal the chief (or senior) (equipped) warrior'; Tomback (1978): 106, 'surely a rank in the army'; cf Krahmalkov (2000): 185, citing *KAI* 73.1/6 (an inscription, in ninth century. Phoenician characters, on a gold pendant found in a seventh/sixth century burial at Douimès, Carthage, in 1894): 'For Astarte, for Pygmalion. Yadomilk son of Paday, an equipped soldier whom Pygmalion equipped', (though contra Tomback (1978): 105, 'who Pygmalion delivered'). *Hanno rab selos*: Slouszch (1942): 207, insc. 207.1–2; Tomback (1978): 299 ff.

67 Polyb. 3.104.4; Livy 22.28; Plut. *Fab. Max.* 11; Rawlings (2016): 215. One wonders whether the men who each selected ten others for the ambush at Trebbia, had an official label. If not *rab halus* then, perhaps, *rab adir arkt* (r'dr 'rkt), translated by Tomback (1978): 299, as 'chief of crack ('dr = mighty/noble/glorious) troops', based on *KAI* 62.4. However, Krahmalkov (2000): 387 s.v. 'RKT, translates this inscription as 'in the magistracy (r) of the prefect ('dr) of public works ('rkt)'.

68 Siebold (2007); Cowdrey (1995); Salo (2011): 87–100 lays out the qualities leaders should possess or develop to enhance vertical bonding; 32–3 defines affective/instrumental components.

69 Wesbrook (1980); Salo (2011): 30.

70 The Numidians already had earlier suspicions of Hanno's bad faith towards their officer, when he risked battle at the Himera while Muttines was absent; at that time, ten Numidians had even approached the Roman general Marcellus to inform him of this dissatisfaction, Livy 25.41.

71 The disastrous negotiations about pay and bonuses in 241 BC notwithstanding; clearly the men suspected the Carthaginian authorities of *not* honouring the promises and commitments of generals, Polyb. 1.67.12.

72 Polyb. 3.80–1: 'Hannibal, therefore, made every provision for carefully attending to the men and the horses likewise until they were restored in body and spirit' after crossing the Alps; cf. 9.22; 11.19; 15.16. Note that prior to Trebbia, 'oil had been distributed amongst the [Carthaginian] maniples for them to make their joints and limbs supple' for the coming wintery battle, Livy 21.55. After Trasimene, he directed his march into Picenum in order to collect supplies that would help his men and horses recuperate, Polyb. 3.87–8.

73 Except in the very last stages of the war when famine and disease appear to have contributed to the enervation of his army in Bruttium; Hoyos (2003): 128–9; Rawlings (2016): 223. Hannibal never appears to have been accused of failing his men. However, clear disregard for the men might result in shame. After Himilco, whose force was consumed by plague at Syracuse in 396 BC, abandoned most of his army to its fate, 'reproaches were heaped upon him' when he returned to Carthage, Diod. Sic. 14.75.4, 76.3.

74 Henderson (1985): 15; Salo (2011): 32–3.

75 These changes involved the locations of encampment, in terrain that would better protect the men by allowing the unfettered deployment of the Carthaginian cavalry and elephants, Polyb. 1.32.2, 4.

76 Rawlings (2018): 159, but note the possible dangers discussed at 167.

77 Rawlings (2010): 284.

78 Lazenby (1996); Hoyos (2000).

79 Rawlings (2007): 13.

80 Some examples of commanders and their men who deserted or defected include: Moericus (Livy 25.29–31, 26.22); Indibilis and Mandonius (Livy 27.17); Attenes (Livy 28.15); Cerdubelus (Livy 28.20).

81 Rawlings (1996): 81–2; Rawlings (2018): 171–2.

82 Cf. Gsell (1928): 392 n. 2.

83 This military law was unknown except a prohibition on alcohol consumption mentioned by Plato (*Leg.* 674a4), which itself was not always observed, especially by its various mercenary contingents; Daly (2002): 83; Gsell (1928): 346. While Roman discipline was admired by Polybius (6.37–8), judging by the contrast he is drawing with Greek practices, it was seemingly also unusual.

84 Dantas (2017): 147–8; cf. Xenophon *An.* 3.4.46–49, where he dismounts after criticism from an infantryman, Wheeler (2002): 141.

Bibliography

Ameling, W. 1993: *Karthago: Studien zu Militär, Staat und Gesellschaft*, Munich.

Armstrong, J. 2016: 'The Ties that Bind: Military Cohesion in Archaic Rome'. In J. Armstrong (ed.), Circum Mare*: Themes in Ancient Warfare* (*Mnemosyne*, Supplement 388), Leiden: 101–119.

Brizzi, G. 1995: 'L'Armée et la guerre'. In V. Krings (ed.), *La civilisation phénicienne et punique. Manuel de recherche*, Leiden: 303–315.

Champion, C. B. 2004: *Cultural Politics in Polybius' Histories*, Berkeley.

Clackson, J. 2012: 'Oscan in Sicily'. In O. Tribulato (ed.), *Language and Linguistic Contact in Ancient Sicily*, Cambridge: 132–148.
Cowdrey, C. B. 1995: *Vertical and Horizontal Cohesion: Combat Effectiveness and the Problem of Manpower Turbulence*, Fort Leavenworth.
Cox, A. A. 1996: *Unit Cohesion and Morale in Combat: Survival in a Culturally and Racially Heterogeneous Environment*, Fort Leavenworth.
Daly, G. 2002: *Cannae: The Experience of Battle in the Second Punic War*, London.
Dantas, D. 2017: 'Xanthippus of Lacedaemonia: A foreign commander in the army of Carthage', *Cadmo: Revista de História Antigua* 26: 141–159.
Donner, H. and W. Röllig, 1968: *Kanaanäische und aramäische Inscriften* (*KAI*), 3 volumes, Wiesbaden.
Eckstein, A. M. 1995: *Moral Vision in the Histories of Polybius*, Berkeley.
Etzioni, A. 1975: *A Comparative Analysis of Complex Organizations*, New York.
Fariselli, A. C. 2002: *I mercenari di Cartagine* (*Rivista di studi punici* 1), Sarzana.
Frey-Kupper, S. 2014: 'Coins and Their Use in the Punic Mediterranean: Case Studies from Carthage to Italy from the Fourth to the First Century BCE'. In J. Crawley Quinn and N. C. Vella (eds.), *The Punic Mediterranean: Identities and Identification from Phoenician Settlement to Roman Rule*, Cambridge: 76–110.
Gat, A. 2006: *War in Human Civilization*, Oxford.
Goldsworthy, A. K. 2000: *The Punic Wars* (repr. as *The Fall of Carthage*), London.
Gruen, E. 1992: *Culture and Identity in Republican Rome*, Berkeley.
Gsell, S. 1928: *L'Histoire ancienne de l'Afrique du Nord, vol. 2: L'État carthaginois*. 3rd edn., Paris.
Hall, J. R. 2018: 'As They Were Ripped from the Altars: Civilians, Sacrilege and Classical Greek Siege Warfare'. In A. Dowdall and J. Horne (eds.), *Civilians Under Siege from Sarajevo to Troy*, London: 185–206.
Hall, J. R. Forthcoming: *Carthage at War: Punic Armies c. 814–146 BC*, South Barnsley.
Henderson, W. D. 1985: *Cohesion: The Human Element in Combat*, Washington DC.
Hoyos, B. D. 2000: 'Maharbal's Bon Mot: Authenticity and Survival', *Classical Quarterly*, New Series 50: 610–614.
Hoyos, B. D. 2003: *Hannibal's Dynasty: Power and politics in the western Mediterranean, 247–183 BC*, London.
Hoyos, B. D. 2007: *Truceless War: Carthage's Fight for Survival, 241 to 327 BC*, Leiden.
Hoyos, B. D. 2010: *The Carthaginians*, London.
Isayev, E. 2017: *Migration, Mobility and Place in Ancient Italy*, Cambridge.
Jenkins, G. K. 1971–1978: 'Coins of Punic Sicily', *Schweizerische numismatische Rundschau*: Part I, 50: 25–78; Part II, 53: 23–41; Part III, 56: 5–65; Part IV, 57: 5–68.
Krahmalkov, C. 2000: *Phoenician-Punic Dictionary* (*Orientalia Lovaniensia Analecta*, 90), Leuven.
Lazenby, J. F. 1996: *The First Punic War: A Military History*, London.
Lee, I. 2000: 'Entella: The silver coinage of the Campanian mercenaries and the site of the first Carthaginian mint 410–409 BC', *The Numismatic Chronicle* 160: 1–66.
Likert, R. 1961: *New Patterns of Management*, New York.
Loreto, L. 1995: *La Grande Insurrezione Libica contro Cartagine del 241–237 A.C.: Una storia politica e militare (Collection de l'Ecole Française de Rome, 211)*, Rome.
MacCoun, R. J. and W. M. Hix, 2010: 'Unit Cohesion in Military Performance'. In B. D. Rostker et al. (eds.), *Sexual Orientation and U.S. Military Personnel Policy: An Update of RAND's 1993 Study*, Santa Monica: 137–165.

MacDonald, E. 2015: *Hannibal: A Hellenistic Life*, New Haven.
Machiavelli, N. 1999: *The Prince*, translated by G. Bull, London.
Mälkki, J. 2014: 'The Linch-pin and the Effectiveness of the Military Organisation'. In A.-M. Huhtinen, N. Kotilainen, and M. Vuorinen (eds.), *Binaries in Battle: Representations of Division and Conflict*, Newcastle upon Tyne: 226–254.
Mildenberg, L. 1989: 'Punic Coinage on the Eve of the First War against Rome. A Reconsideration'. In H. Devijer and E. Lipinski (eds.), *Punic Wars, Orientalia Lovaiensia Analecta 33, Studia Phoenicia 10*, Leuven: 5–14.
Miles, R. 2010: *Carthage Must Be Destroyed: The Rise and Fall of an Ancient Civilization*, London.
Pfeilschifter, R. 2007: 'The allies in the Republican army and the Romanisation of Italy'. In R. Roth and J. Keller (eds.), *Roman by Integration: Dimensions of Group Identity in Material Culture and Text*, *JRA* Suppl. 66: 27–42.
Prag, J. 2010: 'Siculo-Punic Coinage and Siculo-Punic Interactions', *Bollettino di Archeologia on line*, volume speciale A/A2/2.
Pritchett, W. K. 1974: *The Greek State at War*, vol. 2, Berkeley.
Rawlings, L. 1996: 'Celts, Spaniards and Samnites: Warriors in a soldiers' war'. In T. Cornell, B. Rankov and P. Sabin (eds.), *The Second Punic War: A Reappraisal*, *BICS* Suppl. 67, London: 81–95.
Rawlings, L. 2007: 'Hannibal the Cannibal? Polybius on Barcid atrocities', *Cardiff Historical Papers* 9, Cardiff: Cardiff School of History and Archaeology: 1–30.
Rawlings, L. 2010: 'The Carthaginian Navy: Questions and Assumptions'. In G. Fagan and M. Trundle (eds.), *New Perspectives on Ancient Warfare*, Leiden: 253–287.
Rawlings, L. 2016: 'The Significance of Insignificant Engagements: Irregular warfare during the Punic wars'. In J. Armstrong (ed.), Circum Mare*: Themes in Ancient Warfare* (*Mnemosyne*, Supplement 388), Leiden, 204–234.
Rawlings, L. 2018: 'Warlords, Carthage and the Limits of Hegemony'. In T. Ñaco del Hoyo and F. López-Sánchez (eds.), *Multipolarity and Warlordism in the Ancient Mediterranean*, Leiden: 151–180.
Rochette, B. 2012: 'Sur le bilinguisme dans les armées d'Hannibal', *Les Études Classiques* 65 (2): 153–159.
Rosenstein, N. 2012: 'Integration and Armies in the Middle Republic'. In S. T. Roselaar (ed.), *Processes of Integration and Identity Formation in the Roman Republic*, Leiden: 85–104.
Rzepka, J. 2009: 'The Aetolian elite warriors and fifth-century roots of the Hellenistic Confederacy', *AKME* 4: 3–34.
Salo, M. 2011: *United We Stand – Divided We Fall: A Standard Model of Unit Cohesion*, Helsinki.
Siebold, G. L. 2007: 'The Essence of Military Group Cohesion', *Armed Forces & Society* 33 (2): 286–295.
Slouszch, N. 1942: *Thesaurus of Phoenician Inscriptions*, Tel Aviv.
Smith, R. B. 1983: 'Why Soldiers Fight. Part I: Leadership, Cohesion and Fighter Spirit', *Quality and Quantity* 18: 1–32.
Tagliamonte, G. 1994: *I Figli di Marte. Mobilità, mercenari e mercenariato italici in Magna Grecia e Sicilia*, Rome.
Tomback, R. S. 1978: *A Comparative Semitic Lexicon of the Phoenician and Punic Languages* (Society of Biblical Literature. Dissertation Series 32), Missoula.
Tritle, L. 1989: '*Epilektoi* at Athens', *AHB* 3: 54–59.

Trundle, M. 2004: *Greek Mercenaries: from the Late Archaic Period to Alexander*, London.
Walbank, F. W. 1957: *A Historical Commentary on Polybius*, vol. 1, Oxford.
Wesbrook, S. D. 1980: 'The Potential for Military Disintegration'. In S. Sarkesian (ed.), *Combat Effectiveness: Cohesion, Stress and the Volunteer Military*, London: 247–252.
Wheeler, E. L. 2002: 'The General as Hoplite'. In V. D. Hanson (ed.), *Hoplites: The Classical Greek Battle Experience*, London: 121–170.
Whittaker, C. R. 1978: 'Carthaginian Imperialism in the Fifth and Fourth Centuries'. In P. D. A. Garnsey and C. R. Whittaker (eds.), *Imperialism in the Ancient World*, Cambridge: 59–90.
Wollner, B. 1987: *Die Kompetenzen der karthagischen Feldherrn*, Frankfurt.
Zambon, E. 2008: *Tradition and Innovation. Sicily between Hellenism and Rome*, Stuttgart.

6 Roman Standards and Trumpets as Implements of Cohesion in Battle

Adam O. Anders

A fundamental aspect of cohesion in any battle is the means by which soldiers can make sense of their reality. Battle, as a situation with a multitude of constantly changing variables, can make human behaviour unpredictable, particularly as a response to chaos or uncertainty. This, in turn, can lead to disorder, and in the worst cases, the destruction of the unit. To avert this, there are numerous methods of engendering and maintaining cohesion, many of which occur outside the period of battle itself. As Marshall has pointed out, however, during battle, communication between all members of a unit is critical to the unit's cohesion.[1]

Studies on the battle tactics of the Roman army in the late Republic and early Empire, have commonly focused on mass manoeuvres, cause-and-effect analyses, and studies related to the 'face of battle' approach. All these methods gloss over the communication processes in a battle and by extension, how cohesion was maintained. Part of this omission is, of course, due to that most exciting and hypothesis-generating aspect of modern scholarship in ancient history: a dearth of evidence. Yet, ancient military historians, like the men they commit their careers to studying, bravely fight on with what tools they have. As regards unit cohesion in the Roman army, the trifling evidence we do have on the communication process provides us with noteworthy possibilities.

Cohesion on the battlefield may be understood to have two aspects: horizontal cohesion (focusing in this analysis on 'task' rather than 'social' aspects) where a unit performs cohesively as per the commands of a unit leader (e.g., the centurion seeks to coordinate his unit in executing a given order or 'task'), and vertical cohesion, where all units under the command of a general or commander act cohesively as per orders transmitted to all of them. Cohesion, as it is classically defined, is

> … the bonding together of members of a unit or organization in such a way as to sustain their will and commitment to each other, their unit, and the mission. It … enables it to function as single unified, integrated unit. Cohesion allows teamwork to occur when the going gets tough.[2]

DOI: 10.4324/9781315171753-7

Battlefield communication practices, such as aural and visual signals, enhance both vertical and horizontal cohesion, and were a standard part of Roman tactical practice. This chapter will focus on how such communication was executed at the level of the century with a view to clarify our understanding of task cohesion in the Roman army. Some of these practices are applicable to vertical cohesion as well, but discussion will focus on task cohesion since there is more evidence for it.

The success or failure of cohesion has major consequences. This still stands true today, as the failure of vertical cohesion, (i.e., the transmission of commands from commanders to units) in Afghanistan has led to 'friendly fire' disasters.[3] To highlight this point with an example that can be compared to our period, an episode from the Third Crusade (the battle of Arsuf, AD 1191) exemplifies the consequences of the failure of the transmission of commands.

> It was arranged by common consent that six trumpets should be blown in three different places in the army when they were going to engage the Turks, i.e., two trumpets at the front of the army, two at the rear and two in the middle of the army. This was so that the Christians' signal could be distinguished from the Saracens', and to mark the difference between them both during the engagement. If this had been observed all those Turks would have been intercepted and routed, but because the aforesaid knights were in too much of a hurry the general agreement was not observed, which marred the success of the common enterprise.[4]

This use of trumpets is a method of transmitting commands that can be seen in the Roman period as well. The importance of acoustical signals in battle such as these (as well as visual signals) cannot be overstated, and before radios existed, coordinating the movements of thousands of men on a battlefield several miles wide, would have required a well-organized system of transmitting commands.[5]

To clarify our current models of Roman infantry combat and unit cohesion, we require an assessment of what implements were used to engender structure and sensemaking for the soldiers in a century. As Koon has pointed out,

> It is of critical importance … for any model of combat to offer a convincing explanation of how heavy infantry could move backwards and forwards over hundreds of yards while maintaining some semblance of order and formation.[6]

Although scholars are aware of some of the basic means of transmitting commands, this paper will seek to explore these varying systems in greater detail, focusing on the Roman legions, for which we have the most evidence. Through understanding these processes, we may acquire a better understanding of cohesion within Roman centuries on the battlefield.

G. H. Donaldson claims, 'The Roman's use of acoustical and visual signals to control units in combat is well attested'.[7] However, the two pages in Webster's work that serve as a reference to this comment leave much to be desired.[8] While Webster uses iconographic evidence to point out that acoustical and visual signals were used together,[9] his analysis on how they were used does not go much further than this.[10]

Acoustical and visual signals do indeed play a prominent role in the transmission of commands in battle. They are vital for both unit cohesion and vertical cohesion. In his section on military music, Vegetius asserts:

> The trumpet sounds the charge and the retreat. The cornets are used only to regulate the motions of the standards ... in time of action, the trumpets and cornets sound together ... The cornets sound whenever the standards are to be struck or planted. These rules must be punctually observed in all exercises and reviews so that the soldiers may be ready to obey them in action without hesitation according to the general's orders either to charge or halt, to pursue the enemy or to retire.[11]

Dio describes communication at the beginning of battle at the end of the Republic as follows:

> Thereupon watchwords were going around—for the followers of Brutus it was 'Liberty' and for the other side whatever the word which was given out,— and then one trumpeter on each side sounded the first note, after which the rest joined in, first those who sounded the 'at rest' and the 'ready' signals on their trumpets while standing in a kind of circular space, and then the others who were to rouse the spirit of the soldiers and incite them to the onset. Then there was suddenly a great silence, and after waiting a little the leaders uttered a piercing shout and the lines on both sides joined in.[12]

So, Dio indicates that both trumpets and voice commands had a significant role to play in controlling units. The passage from Dio also suggests that the former was used for unit cohesion, while the latter appears as a part of vertical cohesion. Let us assess the usage of sound and instruments in further detail.

Musical Instruments

As their use well into the gunpowder period suggests, horned instruments were the most effective means of transferring commands due to their high-pitch and carrying power, even though several horns and a repetition of the notes may have been necessary to cover the battle-line enclosing possibly several miles in frontage.[13] In the Roman army, musical instruments were a key part of transmitting commands both horizontally (i.e., within the unit) and vertically (i.e., from general to the units) – a fact highlighted by the number of *aeneatores* in each legion. It is necessary at this point to digress slightly on the organization of the army, in order to clarify how the *aeneatores* were distributed amongst the legion, and thus how many men were to adhere to tactical signals given by musical instruments.

In the so-called Servian system (i.e., the organization of the military from about the sixth century BCE to the mid-Republic), there were two centuries of musicians, or approximately 160 men. There is no evidence to indicate that this number changed during the middle Republic, rather, it suits the organization of the Manipular legion perfectly.[14] With 80 centuries in the Manipular army, we have one *tubicen* and *cornicen* per century. For the imperial period the amount of musicians seem to have changed, at least according to the list of trumpeters in *legio* VIII *Augusta* at Lambaesis: there are 73 (35 *tubicines* and 38 *cornicines*).[15] Applying this number accurately to an imperial legion, as Speidel put it, 'is a matter of hopeless conjecture'.[16] Regardless, there is no reason not to assume that on paper, there were to be two trumpeters per century, and indeed we have epigraphic evidence that suggests as much for the imperial period.[17] Keeping this in mind, we may return to discussing the significance of *aeneatores* and their role in cohesion.

Most of the literary references to musical instruments refer to the sounding of the charge and the retreat[18] – confirming Vegetius' assertion. Indeed, maintaining vertical cohesion during a charge or retreat would not have been possible without a means of transmitting the command at great volume, over an extended distance: something only instruments could do most efficiently and quickly.[19]

Donaldson evaluates any possible relationship between the military musicians (*aeneatores*) and the *immunes* of the army, to determine whether they were specialists, who could possibly play complicated sets of notes that might signify a wide variety of commands. Although there is one inscription referring to a *tubicen* and *cornicen* registered as *immunes*,[20] Donaldson asserts that this is an exception; and argues that it is most likely that any soldier could be instructed to use the instruments, concluding that they could only blow simple calls with them.[21] However, while it may be true that *aeneatores* were taken from among the poorest citizens during the Republic,[22] Josephus' mention of their association with the sanctity of the eagles (on the march in particular) indicates they were held in higher regard during the Principate.[23] Breeze and Speidel also affirm that the *aeneatores* were specialists within the legion, due to their status as privileged officers (*beneficiarii*), as indicated in the epigraphic evidence; as well as their need for special coverage in battle, their knowledge of the command process and/or their leadership ability.[24] Furthermore, Donaldson's view that only the simplest acoustical signals would have been used in battle is somewhat misleading. Single note commands seem unlikely, as they might be hard to discern. Indeed, modern reconstructions have provided players with the ability to play up to eight different notes in some cases,[25] and Vegetius states that the instruments signalled the army to advance, retreat, halt, pursue fugitives or to be recalled.[26] This series of notes was produced without keys or holes, by the modification of the breath and the lips at the mouthpiece;[27] so that to produce at least five different notes for the aforementioned commands would likely have required some training, and so, not just any soldier would have been able play them, as Donaldson suggests.[28] This specialization of the *aeneator* is most significantly articulated in our sources by Aristides Quintilianus, who says that Rome

> often rejects verbal orders as damaging, if they should be discerned by those of the enemy speaking the same language, and makes codes through music by playing the *salpinx* – a warlike and terrifying instrument – and appointing a specific *melos* for each command. When the attack was by line and the approach was by column, she set down special tunes, and a different kind for retreat; and when the pivoting was to the left or to the right, again there were specific tunes for each; and so she accomplishes every manoeuvre one after another by means of codes that are on the one hand unclear to the enemy and on the other hand both totally clear and easily recognized by the allies. For they do not hear these codes only in part, rather the whole corps follows a single sound.[29]

The advantages of transmitting various commands through horn signals rather than by verbal orders (as was the case for Greek armies),[30] are perhaps obvious, but Quintilianus' explanation suggests that every specific tactical movement was coordinated by trumpet.[31] This suggestion has significant consequences for an investigation into task cohesion and so it worth examining further.

The likelihood of Quintilianus' suggestion may be judged by comparing the number of specific tactical movement commands we are aware of. The *Oxford Classical Dictionary* lists 12 commands that use the word *signa*, and if we take Vegetius' suggestion that the standards and horns worked together to be true, then it seems likely that the horns also had a good number of various tunes, according to the tactical movement required. We will return to the coordinated issuing of orders between horn and standard below.

Iconographic evidence further reveals the possible usage of instruments in the transmission of commands. As Webster points out, scene 106 on Trajan's column depicts a centurion turning to the *cornicen*, suggesting he is giving the order to blast a note, so that attention might be drawn to the standard. Although the scene depicts the army is on the march, what is most important to note is the idea that the trumpets are used in collaboration with the standards, which coincides with Vegetius' suggestion that the sound of the trumpet was to draw attention to them.[32] This can also be attested elsewhere on Trajan's column. Scene 41 shows the Romans in battle, and the *cornicens* dutifully standing by the standards. We will return to this connection in more detail later.[33]

Evidence that musical instruments played a significant role as accoutrements of battle can also be seen in the iconography of victory. Representations of war booty and equipment-trophies include the depictions of musical instruments, such as those found on the base of Trajan's column. Also, in his recounting of the battle of Vercellae, Plutarch mentions the trumpets being part of the war booty.[34] Instruments may have been valued for their metal, which could have been reworked to suit the victorious army's needs, or kept as victory trophies, as attested above. The general prominence of musical instruments in military iconography and in the literary sources clearly indicates that they must have played a significant role in battle. Their value as war booty, therefore, must have also been based on their significance in transmitting commands and in cohesion; depriving an enemy of his means of transmitting commands was a serious blow and a major tactical advantage for the winning side in a longer campaign. It is worth examining then, at this point, how the trumpets affect cohesion.

The variety of messages discussed above is directly linked to a range of possible situational realities. What is important to draw from this is the trumpet's role as an implement of sensemaking (to employ a useful term in relation to cohesion) for the soldiers.[35] It creates an objective understanding of the task in which the soldiers are involved. With the ability to play different notes, it can tell soldiers what the status of the task is in simple terms – whether it is succeeding as planned, or whether it requires alternative action. It thereby constructs reality for the soldiers in a situation where objective reality cannot always be grasped. The trumpet thus 'make sense' of the mission and the danger, providing a sense of order and rationality that can be grasped.

Accordingly, their main role in battle was ensuring unit cohesion, and thus control. They were an extension of the commander: a proxy for front-line generals. In this way, depicting instruments and standards in victory iconography can be interpreted to be symbolic of having removed the implements of sensemaking from the enemy, thus causing incoherency or the disintegration and utter defeat of the enemy, its commander, and any associated prestige he may have had.

As has been suggested, the trumpets of the army may have had a close relationship to the standards in transmitting commands. As such, their role as implements of sensemaking should also be examined.

The Military Standards

In order to clarify the role of standards, it is advisable to begin with some discourse on both why they were important to the Romans (and to armies in general up until the modern period), beyond their primary function as implements of both vertical and horizontal cohesion.

One of the advantages Rome had in her army was that the tactical unit was a small one and each of these centuries had a standard.[36] The standard served as a rallying point, a symbol and cause of esprit de corps, and as an identifier of its unit.[37] They were the focus of the unit's pride, and so they were highly valued; losing it would not only disturb a unit's ability to position itself properly in a battle, but it would also mean the loss of the unit's identity. In short, they were the literal and figurative objects of unit cohesion. Because the standards also held a sacred significance, their loss called for serious punishment to the bearer who did not defend it with his life.[38] Indeed, they were defended vigorously: at Jerusalem in 70 CE three Roman standards were captured only after vicious fighting,

> ... those that bore the ensigns fought hard for them, as deeming it a terrible thing, and what would tend to their great shame, if they permitted them to be stolen away.[39]

The value of standards has persisted even through the Napoleonic Wars, despite the fact that they did not have the religious significance for the French army (and its opponents) as they did for the Romans. The capture of the eagle from Napoleon's 45th regiment by the Royal Scots Greys at Waterloo was the event that gave the unit its current badge: the symbol of the eagle.[40]

The main types of standards may be categorized into *signa*, *vexilla*, *imagines*, and *aquilae*. Both *signa* and *vexilla* are applied to standards belonging to a unit within the legion, and so they will be the focus of the discussion.[41] They also may be discussed together. Because these standards created cohesion both through their symbolic significance as the spirit of the unit they represented, as well as tactically through their visibility on the battlefield, it may be worth first reviewing how they looked.

The *vexilla* was a standard with a hanging fringed piece of cloth, approximately 50 centimetres square on average, placed on a pole that may have been up to 3.60 meters in height. It likely had the unit's name (e.g., century, cohort, legion during the Late Republic/Empire) and possibly a symbol of the unit.[42] These were the oldest standards of the Roman army, and, after Augustus, their usage may have changed from being standards for each unit to standards for divisions of infantry that were separated from the main body of troops for some special duty. The *signa* were poles along the top of which there was a transverse bar where a plate was sometimes placed containing the name of the legion, cohort, and century. Below the transverse bar came a series of discs, probably of silver, like the *phalerae*, followed by a crescent moon (probably an amulet to avoid bad luck) which may have been preceded by the unit's symbol in some cases.[43] If a legion or unit had gained any special distinction, that ornament was affixed to the pole of their standard.[44]

Above the bar, there may have been a *corona aurea,* usually accompanied by an open palm (the token of fidelity). These may have been the standards to replace the *vexilla* under Augustus.[45]

All the standards had a point at the lower end for fixing into the ground, as well as handles to remove them afterwards.[46] These were important design characteristics that helped a standard-bearer participate in battle and defend the all-important standard.

Iconographic and literary sources often suggest ways in which standards were used in battle. This evidence will aid in our understanding the role standards played in unit cohesion. A scene from Trajan's column indicates that the standards held a prominent place in combat, holding their position just behind the front lines. Here they would have served as rallying points,[47] as well as a guide for the soldiers in combat.[48] One of the best examples of this is in the account of the African War: Caesar's troops becoming disordered are ordered not to advance more than four feet beyond the standards. Soon after, the soldiers are surrounded and so in an attempt to extend his line, Caesar has every other cohort press to the rear of the standards, while the remaining cohorts press to the front of them.[49] Not having standards or a disordered movement of the standards confuses soldiers and leads to lack of order in the fighting, as witnessed by Caesar's soldiers when they were ambushed by the Nervii and when Q. Cicero's camp was attacked by the Germans.[50] In the Jugurthine War, the Romans, 'could not keep their ranks' when they were unable to follow the standards, and so some soldiers gave way, while others tried to attack.[51] Standards were vital for controlled troop movements, even for barbarian armies.[52] Thus, since taking a standard prevented a unit from functioning effectively in battle, this must have been a major reason that standards were greatly valued as war booty. Just as with taking an enemy's musical instruments, it interfered with the transmission of commands or disrupted the spatial coherence of the unit (by denying the visual cue for formation/rallying) and so disrupted unit cohesion.[53] The soldiers' successful response to the standards and horns reinforced the importance and effectiveness of these communication tools. Success was naturally a reinforcer of unit esprit de corps as well. In this way, every new task assigned to a unit increased its cohesion. This goes some way to explain the sacred nature of the standards. Soldiers' success, and, by extension, lives, were intimately tied to their standards. In the same ways the gods were said to control human fate, the standards were also a non-human entity that controlled soldiers' fates. Ignoring or losing touch with them could have been fatal – as with the gods. Perhaps this similarity then, had some role in giving the standards sacred value.

The above evidence makes it clear that the standards had a prominent role to play in transmitting tactical commands to the soldiers in battle. We are also made aware of their strategic significance by the number of tactical phrases that use the word *signa.*[54] They determined where the troops were positioned, which was a commander's foremost concern once the battle was underway.

Like trumpets then, standards were implements of sensemaking for the soldiers. The variety of tactical phrases that use the word *signa* suggests, as with the various notes on trumpets, that the standards were meant to relay information to the soldiers on a variety of possible realities or orders of procedure. This ability to communicate a broader sense of reality reflects the standards' role as conveyers of objective reality for the soldiers. Furthermore, as we have seen, the removal of this implement of sensemaking often leads to incoherency or the lack of cohesion, frequently resulting in the destruction of the unit. It creates what Weick has termed a 'cosmology episode'. As Weick explains

> A cosmology episode occurs when people suddenly and deeply feel that the universe is no longer a rational, orderly system. What makes such an episode so shattering is that both the sense of what is occurring and the means to rebuild that sense collapse together. Stated more informally, a cosmology episode feels like *vu jàdé*—the opposite of *déjà vu*: I've never been here before, I have no idea where I am, and I have no idea who can help me.[55]

Such episodes lead to the dissolution of cohesion in the Roman army, particularly when standards are lost, as seen at the Battle of the Sabis.[56] In this battle, the standards became bunched together, and thus incapable of functioning in their normal role as sense-makers. The same happened at the Teutoburg forest, where the lack of space created the ultimate cosmology episode resulting in the infamous Varian disaster.[57] Importantly, in the Teutoburg forest, the trumpets played a role in saving some of the soldiers.[58] Both standards and trumpets are therefore crucial in making sense of reality for soldiers in battle and indicating to them what must be done. Standards and trumpets are thereby primarily responsible for unit cohesion. Simply stated, standards and trumpets, as implements of sensemaking, are fundamental implements of cohesion in combat.

The aforementioned iconographic and literary evidence suggests that auditory and visual cues may have worked together in the command process. The following section will aim to determine the extent to which this was true, and how it positively affected Roman unit cohesion.

Sound and Visuals in Battle: Implements of Horizontal and Vertical Cohesion

As we have seen above, our sources suggest that the musical instruments could give various tactical orders, and we also know that there were probably at least 12 oral commands related to how the standards should direct troops. Whether the latter was articulated by a physical movement of the standard is not always certain, and any speculation on how the standards would have moved in order to express, for example, a pivoting movement, is baseless. Moreover, such subtle movements could be misinterpreted if, for example, a strong wind was affecting the standard. Thus, the standard itself, even if accompanied by an oral command, may not have been enough to issue clear orders to men engaged in combat. Accordingly, the horn could make up for this shortcoming. So, based on the earlier evidence that musical instruments gave tactical signals to troops, I suggest it would have been necessary for the standards to respond to these orders, moving in a way that would correspond to the tactical command signified by the note or notes played. This process could have worked vice versa as well, with a unit's *cornicen* or *tubicen* musically articulating the tactical movement of the nearby standard.

Both audio and visual commands would have been necessary because of the way they complement each other in certain battle conditions. For example, in battles that were fought under poor visual conditions such as heavy fog, dust, or at night,[59] the standards may have been difficult to see. Alternatively, some soldiers may have been too busy fighting to notice the position of the standards.[60] In either case, the sound of the horns would have to guide the positioning of the soldiers in battle. On the other hand, there may have been certain battles where the sound may not have carried as well, since sound carries based on a number of factors such as humidity, wind, temperature, altitude, surrounding terrain, and of course, position of the instrument in the unit.[61] In some cases, sound has been known to bounce unexpectedly, due to a phenomenon known as 'acoustic shadowing'. This was witnessed several times during the American Civil War, when commanders often depended on what they could hear of the battle to tell them what was going on at their forward units. At the Battle of Gettysburg, such acoustic shadows may have been the cause of the confusion of the Confederate attack and their eventual loss.[62] If an acoustic shadow occurred on a Roman battlefield, it may have prevented successful vertical cohesion such as a coordinated attack or the initial charge, signalled by the horns that repeat the command down the line. Another event that would cause a unit to lose their acoustical command signals would be if the *aeneator* had been killed in combat.[63] In such situations where the sound of the tactical commands relayed by the instruments could not be heard, the standards would have to take one the sole responsibility of transmitting directives to the troops. If the standards were approximately 3.60 meters in height as Rostovtzeff suggests,[64] they would have been easily visible to troops looking to them for guidance (in decent visual conditions). Nevertheless, both the standards and the horns complimented each other, and would normally have had to work together, so as not to give conflicting commands.[65] Since these tools were an extension of the general (by visually and audibly transmitting his commands – reminding them of his continual 'presence' in the battle), it should come as no surprise that soldiers felt lost or demoralized without these implements.

The Position of the Standards and *Aeneatores*

Based on the evidence discussed so far, it is likely that the standards and the musical instruments worked in close proximity of each other. This in turn raises the question of where they were positioned in the battle line since this would give us a clearer picture of how they worked in practice as implements of cohesion. Domaszewski has argued that the standard-bearers occupied the front line of battle.[66] Though they could have functioned as a proxy for front-line generals, the very front of a unit would be an extremely dangerous position for such a vital officer in the army, especially since he was not only holding a very valuable object in the enemy's eyes, but also because he was equipped with a smaller shield than other legionaries (thus making him an even more appealing target). He would have been a prime objective, as we have seen from the struggle that the standard bearers faced in Jerusalem.[67] Thus, as Oakley has suggested, it seems likely that the standards were carried a few feet behind the front lines (i.e., somewhere behind the first rank, giving said rank room to manoeuvre).[68] Evidence from Livy, Caesar, and Dionysus[69] support this, as do the depictions on Trajan's column.

While this is a safer position than the front line, it did not mean that the standard-bearers never engaged in combat. The best evidence indicating that they did engage in combat (aside from direct mentions of it, as in Joseph. *BJ* 6.223–6 and others) is the design of the standards themselves. As mentioned above, the standards could be planted in the ground, and then removed with their handles afterwards. This would have given the bearer the necessary freedom to manoeuvre against an enemy in combat. Reference to this planting of the standards is also cited in Livy, where he mentions that the command *signa convellere* was given, i.e., 'to take up the standards' which had been fixed in the ground.[70] The fact that standard-bearers were officers[71] and that they were armed with mail and a sword,[72] indicates that accomplished soldiers were the only ones trusted to have enough experience and/or skill to defend the valuable standards effectively. As to whether the *aeneatores* could have stood in close proximity to the standards as I have argued, I would point out that the *cornicines* on Trajan's column are all depicted as being fully armed, their armament is the same as the standard-bearers'. This suggests to me that they would have been threatened by the danger of combat in the same way the standard-bearers would have been. If a *cornicen* was in mortal danger, he could presumably throw down his instrument (unless honour and discipline prohibited it), and defend himself with his sword, and if his instrument were destroyed in the process, then the unit would rely on the standards for tactical commands, as I have argued above. Thus, it is probable that Vegetius' suggestion regarding the trumpets and the standards working together is correct.

Having clarified the practical uses of these implements of sensemaking for soldiers, it is worth examining some theory behind their cohesive effects.

A Model of Cohesion Components

One of the standard models of military group cohesion is proffered by Siebold as follows, it

> consists of four related, interacting components based on different structural relationships: peer (horizontal), leader (vertical), organizational, and institutional bonding. Peer or horizontal bonding is among members at the same military hierarchical level (e.g., squad or group members). Leader or vertical bonding is between those at different levels (e.g., between squad or group members and their leaders). Peer and leader bonding within a small group (e.g., a platoon) together compose primary group cohesion.[73]

As this chapter has focused on the role of standards and trumpets in battle, Siebold's definition of primary group cohesion is most useful here. The aforementioned implements of sensemaking can be said to create primary group cohesion both symbolically and practically in battle. Siebold has also argued that the essence of strong primary group cohesion is trust among group members and a capacity for teamwork on a given task.[74] Because the standards and trumpets function as implements of sensemaking for soldiers in combat, soldiers' capacity for teamwork is heavily reliant on these tools. Standards and trumpets thus maintain cohesion. To underscore this point, it is worth reviewing Siebold's view of cohesion, which

> is a social-relationship product or form generated by the interactions and experiences of the group members in the context of their daily military activities, combat and noncombat.[75]

The trumpets and standards of the Roman army may be seen as a tool for engendering this social-relationship product, whether in noncombat situations where the men identify on a personal level with the unit that the standards represent, and necessarily in combat situations where the understanding of their relationship to one another depends on the implements of sensemaking. As King has put it

> Without a common external reference point, individuals could not be certain of the status of their social relations or of their obligations to each other. Symbols actualized the purely ideational concept of group membership.[76]

Standards in the Roman army actualize this concept of group membership. Their visibility and movement, partnered with the sounds of the trumpet, speak to the status of every member (moving) as a whole in battle. In this way, they represent cohesion and are a manifestation of it, while at the same time, creating cohesion. The implements of sensemaking thus create a closed cohesive loop in battle.

How they functioned in the chain of command should also help to clarify their role in unit cohesion.

The Chain of Vertical Cohesion

The successful transmission of commands in battle required a pre-established command structure. In Livy, we find examples of commands being given to the standard-bearers directly; one of these examples is a centurion giving orders, and the other, a general.[77] This highlights the importance the Romans placed on task cohesion: each new task was delivered to the specialist designated to transmit that task to the unit, and as such, even the general might engage with said specialist directly.

Webster, in his note on scene 106 from Trajan's column suggests that the centurion, depicted turning back to the *cornicen*, is commanding the movement of the standards through the *aeneator*.[78] It is a well-known fact that the centurions were expected to lead from the front and by example, which resulted in their uncommonly high casualty rates. Thus, although they may have been primarily concerned with fighting, they could also possibly have shouted commands, to guide the standards when necessary.[79]

Appian describes this process in his description of the beginning of the Battle of Pharsalus:

> So [Pompey] gave the signal first and Caesar re-echoed it. Straightway the trumpets, of which there were many distributed among so great a host, aroused the soldiers with their inspiring blasts, and the standard-bearers and officers put themselves in motion and exhorted their men.
> (App. *B Civ.* 2.78)

Once a battle was underway, different units needed various commands based on their individual situations. In this case, vertical cohesion may have also relied on extra-unit soldiers, i.e., messengers, used to ensure task cohesion of the unit within the entire army. By creating new tasks, vertical cohesion reinforced cohesion, solidifying the link soldiers had to the implements of transmitting commands, as well as those involved in the process, from general to officer.

Messengers were used in battle as a means of communication between the commander-in-chief and his officers.[80] Messengers being sent from the commander, and arriving at the standard bearer, *aeneator*, or centurion to relay the order, could have been a possibility.[81] In the Imperial period it may be argued that there were officers specifically in charge of giving commands to smaller units, such as a century. One of these was the *tesserarius*. As his title suggests (*tessera* being the tablet marked with the watchword), this officer was in charge of the daily watchword. The most common function of the watchword was as a security measure for the camp; however, evidence suggests watchwords were used in battle as well.[82] Although there is no direct evidence attesting to the *tesserarius*' role in battle, there is evidence suggesting that they were entrusted with orders to be passed to their fellow officers.[83] We might surmise that they were responsible for receiving commands accompanied by the watchword as a mark of the command's authenticity. They were probably also responsible for giving and receiving commands from *socii* and *auxilia*, as their identities might have had to be authenticated with the watchword. In civil war battles, the use of watchwords to verify commands was vital.[84] Since there was a *tesserarius* posted with each century, it is further likely that they played a role in transmitting commands to their immediate fellow officers, the centurion, the standard-bearer and the *aeneator*. Although the *optio* is another important officer in each century, I believe their position at the back of the formation signifies their role in keeping the soldiers in line, rather than being involved in the transmission of commands. Nevertheless, their role in unit cohesion should not be dismissed. We should conclude, then, that the *tesserarius* probably had an important role to play in unit cohesion, particularly at the level of the century (i.e., the successful execution of tasks).

It is also known that tribunes had a commanding role in battle; Caesar chastises his for not controlling the soldiers at the Siege of Gergovia.[85] I suggest that the tribunes' communication role was primarily one involving vertical cohesion: they may have controlled the actions of the cohorts of their legion, either by directing the *tesserarii* or the other officers of the unit (the centurion standard-bearer and the *aeneator*). Above the tribunes in the chain of command would have been the legionary legate. There are many examples of the legates and/or the tribunes being put in command of sections of an army,[86] and similar divisions of command are made in foreign armies.[87] They are also the principal officers in attendance in a council of war.[88] Thus, in charge of part of an army, and aware of the general's overall battleplan, it was likely the legates and the tribunes who would have made their subordinates such as the centurions and standard-bearers aware of the tactical orders, and one often finds the legates and tribunes exhorting the legionaries in battle.[89] In the later Republic, the generals themselves acted in a similar way, personally leading men, exhorting them, and sending and signalling orders for tactical movements.[90] Yet, as Caesar makes clear, it was difficult for one man to be in control of an entire army,[91] and on days where the weather or terrain was unfavourable, it would have been impossible to know the disposition of the entire army.[92] This is why the implements of cohesion (i.e., standards, trumpets, and the aforementioned officers) could serve as a proxy for the general and as such were crucial to Roman success.

Conclusions

Vertical cohesion depended on a chain of command. From the top, there was the general, who in the later Republic would have actively given orders during the battle, and sometimes even led and exhorted his men personally. During the Imperial period, the emperor, as commander -in-chief, was much less likely to engage in the combat himself, since his life was highly valued. However, for the command to charge or retreat, he would have had the *cornicines* accompanying him send this command out to the entire army.

Command was also delegated to the legates and the tribunes as the next officers in the chain of command. In charge of a part or wing of the army, the legate and his tribunes may have led and exhorted the legionaries personally, but likely observed tactical movements and sent orders to adjust them, according to the general's plan that they had been made aware of at the council of war. These orders may have then been received by the tribunes if the order was for an entire cohort, and/or if for individual centuries, the *tesserarii*. The *tesserarii* could then relay the current orders to the centurion, standard-bearer, and *aeneator* in their century. A blast on the trumpet and the movement of the standard would be the final link in the transmission of commands that would get the soldiers to move appropriately. Knowing this, the general or the legate, if he needed a quick response could have personally ridden down to the appropriate standard and made the standard-bearer and *aeneator* aware of the necessary commands. Every success attained through this chain reinforced its value, trustworthiness, and probably its efficacy as well. When tasks arriving at, and executed by, the unit were carried out successfully, this likewise worked to build trust between the links in the chain, strengthening its efficacy, as well as esprit de corps. In this way, vertical cohesion reinforced task cohesion.

Standards and trumpets played a crucial role in making sense of the battle for the soldiers, and without them, the chaos of a so-called cosmology episode could ensue, potentially resulting in the destruction of the unit or army. This critical element of Roman military success was not seen again until the high medieval period. Acoustical and visual signals remain important aspects of military operations to this day, and the ability for these signals to be transmitted to the soldiers carrying out a given task is critical to said task's success.

The organization of the Roman army, therefore, had a significant role to play in vertical cohesion, but it was the standards and trumpets that may be counted as the primordial elements of horizontal cohesion.

Notes

1 Marshall (1968): 47.
2 Van Epps (2008): 4.
3 *Guardian Unlimited* (2002), 'Four Canadians killed in 'friendly fire'. Available at: http://www.guardian.co.uk/afghanistan/story/0,1284,686313,00.html Accessed: 19 May 2007. 'On 5 December 2001, a B-52A dropped a bomb on US and allied Afghan forces near Kandahar, killing three Americans and at least seven Afghans … The investigation is not complete, but officials have said there *were errors in transmitting target coordinates* to the plane's crew'.
4 Nicholson (1997): 252.
5 For size of battlefields, see, for example, the battlefield maps in Edwards' translation (1917) of Caesar's *Gallic Wars* (623–5, 628–9, after Colonel Stoffel); calculations of the length of battlelines, Taylor (2014): 310–18, esp. table 2.
6 Koon (2011): 88.
7 Donaldson (1988): 349.

8 Webster (1969): 135–6.
9 See below p.000.
10 A single reference to Caes. *Bell. Af.* is all Webster (1969) provides as evidence for the transmission of commands, aside from the iconography mentioned above, n. 8.
11 Vegetius *Mil.* 2.22: *Tubicen ad bellum uocat milites et rursum receptui canit. Cornicines quotiens canunt, non milites sed signa ad eorum obtemperant nutum. … quotiens antem pugnatur, et tubicines et cornicines pariter canunt … Cum autem mouentur signa aut iam mota figenda sunt, cornicines canunt. Quod ideo in omnibus exercitiis et processionibus custoditur, ut in ipsa pugna facilius obtemperent milites, siue eos pugnare siue stare siue sequi uel redire praeceperint duces.*
12 Dio Cass. 47.43.
13 For a signal being given to an entire battleline at once: Caes. *BG* 7.27, 7.45, 7.62; Plut. *Pomp.* 70; App. *B Civ.* 2.78; c.f. Goldsworthy (1996): 150. For size of battlefields, see above, n. 5.
14 Livy 1.43; Dion. Hal. *Ant. Rom.* 4.16–18; cf. Daly (2002): 50, 60; Cornell (1995): 179.
15 Speidel (1976): 161 esp. n. 122.
16 There is no way to resolve the number of 73 satisfactorily. One suggestion might include two musicians for every two centuries. The result is 54 musicians in the bottom nine cohorts, and then likewise assigning two musicians (one of each type) to each of the five centuries in the doubled first cohort, totaling 20, thus resulting in 74 trumpeters in the legion. Applying such a calculation leaves one musician missing from the legion at Lambaesis, though there could have been any number of reasons for this, including temporary leave, sickness, etc. Thus, we probably have one trumpeter per century in the bottom nine cohorts and two per century in the first cohort. A further issue is that not every century has the same type of trumpeter, so Speidel's assertion is put quite adequately.
17 *CIL* III 7449, cf. Speidel (1976): 155.
18 Sounding the charge: e.g., Caes. *BCiv.* 3.46; Plut. *Pomp.* 70, *Sull.* 14, 29, *Aem.* 33; Suet. *Tib.* 37; Tac. *Ann.* 1.68, 2.81, 4.25; Joseph. *BJ* 3.7.27, 4.1.4, 6.1.7; App. *B Civ.* 2.78, 5.38; Dio Cass. 36.49. Sounding the retreat: Caes. *BG* 7.47; Joseph. *BJ* 2.20.7. Most of the sources simply refer to a *cornu* or σαλπιγξ, commonly translated as 'trumpet', which is the curved horn depicted in the iconographic evidence discussed below. For a description of the individual horns described by Vegetius see Smith *et al.* (1891) s.v. *Cornu*; *Tuba*. The *bucina* and *lituus* were not used in transmitting commands, but rather just for signalling the preparation for battle and other camp functions; thus, they were not assigned to each century like the *cornu* and *tuba*: Speidel (1976): 154, esp. n. 107, and 155; Webster (1969): 142. For a discussion of the different horns, see Speidel (1976): 148–52.
19 E.g., Caes. *Bell. Afr.* 18: *Itaque signo dato … subito immittit cohortes turmasque suorum*, which suggests the entire force charged at about the same time, thus, the 'signal' in this case would have had to have been an acoustical one; cf. n. 13.
20 *CIL* III 7449.
21 Donaldson (1988): 351.
22 Livy 1.43; cf. n. 14.
23 Joseph. *BJ* 3.115, 5.48.
24 Breeze (1969) 52; Speidel (1976): 160.
25 Myers (2006) suggests that an expert musician could get 7–8 notes from the *cornu*, and that only three notes might be easy for anyone to play. Anderson (1970): 80, notes that the Greeks were unable to give more than two calls and this is why verbal orders were always issued prior to the trumpet blast, he goes on to suggest that from this the 'elaborate' system of the Roman imperial army may have developed (1970: 81, esp. n.58; and see below n. 54).

26 Veg. *Mil.* 3.5 *Nam indubitatis per haec sonis agnoscit exercitus, utrum stare uel progredi an certe regredi oporteat (utrum longe persequi fugientes an receptui canere).*
27 C.f. Smith, *et al.* (1891), s.v. *Cornu,* 544.
28 Although the possibility exists that various commands could be executed by playing a single note in different patterns, this too would require some skill, cf. Veg. *Mil.* 3.5: *cornu ... temperatum arte spirituque canentis flatus emittit auditum.* Also, K. Gilliver (pers. comm.) has pointed out to me that in her personal experience, not everyone is able to purse their lips appropriately in order to play a brass instrument.
29 Aristid. Quint. 62.6–19, trans Mathiesen (1983); cf. Krentz (1991): 117.
30 Krentz (1991): 117–18.
31 The Salpinx was key to phalanx combat. It was used to call for silence, to order men in line and call them to arms, sounding the charge and for retreat, see Krentz (1991): 114–16.
32 See n. 11.
33 See 'Sound and Visuals in Battle', below.
34 Plut. *Mar.* 27; see also n. 53.
35 Cf. Van Epps (2008), who argues that sensemaking of reality is critical to cohesion.
36 Small tactical units: Bell (1965): 412; Smith *et al.* (1891): 672. A standard per unit: Smith *et al.* (1891) 672; c.f. Connolly (1998): 216.
37 Lee (1996): 208; as an identifier of a unit, this would have been especially important in battles where soldiers would have been outfitted in similar armour, e.g., during civil wars.
38 Sacred standards: Plin. *HN* 8.23; Haynes (1993): 143. Punishment for losing a standard: Caes. *BCiv.* 3.74; Brand (1968): 156; this is further exemplified in standard-bearers giving their lives in defence of the ensigns, e.g., Livy 34.46; Caes. *BG* 5.37 and below.
39 Joseph. *BJ* 6.223–6.
40 The British Army. *The Royal Scots Dragoon Guards (Carabiniers & Greys).* [online]. Available at: http://www.army.mod.uk/scotsdg/index.htm [Accessed 22 May 2007].
41 For discussion of the *aquila*, see Greet, this volume.
42 Rostovtzeff (1942): 94 *et passim.*
43 Smith *et al.* (1891): 73, amulet: c.f. Plaut. *Epid.* 5.1, 38 (638).
44 Smith *et al.* (1891): 675.
45 Smith *et al.* (1891): 675.
46 Smith *et al.* (1891): 673–5; Connolly (1998): 219.
47 Caes. *BG* 2.21, 2.25, 5.34; *BCiv.* 1.44.
48 Livy 9.13.2, 22.5.3, 33.7.2, 37.39.5.
49 *Bell Afr.* 15 and 17. The verbs used to describe the issuing of these orders are *edicio* and *iubeo*, respectively. Perhaps it is obvious that the verbs themselves do not tell us the exact means by which the orders were transmitted, however, see 'The Chain of Vertical Cohesion', below, for suggestions.
50 *BG* 2.25 and 6.37; see also *BG* 4.26, 5.16.
51 Sall. *Iug.* 51.
52 Caes. *BG* 4.15.
53 Standards as booty: e.g., Caes. *BG* 7.88; Livy 39.31; Plut. *Mar.* 27; Sall. *Iug.* 74.
54 E.g., *signa sequi, signa subsequi, signa observare, signa servare, ferte signa in hostem, signa constituere, signa proferre, signa convertere, signa inferre, signa conferre cum aliquo, signa convellere, signa hostium turbare, ante signa, post signa.* See also *The Oxford Classical Dictionary* (1970) s.v. *signa.*
55 Weick (1993): 633–4.

56 Caes. *BG* 2.25, cohesion was reportedly only restored by Caesar's personal intervention at the front lines.
57 See Dio Cass. 56.21.2, for the lack of space.
58 Dio Cass. 56.22.3.
59 Fog: Cynoscephalae, Polyb. 18.20.7, Livy 33.7; Magnesia, App. *Syr.* 33. Dust: Vercellae, Plut. *Mar.* 26; Carrhae, Plut. *Crass.* 25; Dio Cass. 40.23. Night: Colline Gate, App. *B Civ.* 1.93, Plut. *Sull.* 29; Bibracte, Caes. *BG* 1.26; Alesia, *BG* 7.81; 2nd Bedriacum, Tac. *Hist.* 3.53; note the poor light in App. *Syr.* 33 where 'The day was dark and gloomy so that the sight of the display was obscured and the aim of the missiles of all kinds impaired by the misty and murky atmosphere'.
60 As Lee (1996): 201, points out, the amount of noise in some battles alone could be confusing or intimidating, so much as to distract soldiers from what was happening to the standards behind or beside them. Anderson (1970): 79 argues that this was the reason for the trumpet blast in Greek warfare, to function as an 'executive order': the verbal order came first, and the trumpet executed it; cf. Xen. *Hell.* 5.1.9.
61 Regarding where the *aeneator* may have stood in battle, see my argument below.
62 Ross (2001).
63 This raises the question of whether an *aeneator* would engage in combat due to the encumbrance of his instrument, which I have addressed below.
64 Rostovtzeff (1942): 94 *et passim.*
65 cf. Anderson (1970): 79; see above n. 60. Also, note Sparta's loss at Leuctra, when they were not led by their flute players, as usual: Polyaenus *Strat.* 1.10; cf. Anderson (1970): 82.
66 Domaszewski (1967); Smith *et al.* (1891): 672.
67 Joseph. *BJ* 6.223–6.
68 Oakley (1998): 509–10, who indicates that an enemy might advance beyond the standards without capturing them. He also points out that the front line was movable depending on the situation: i.e., it could have an undulating shape, which in turn explains how the enemy can advance beyond the standards in places without them being seized. See also Taylor (2014) for tactical spacing of units.
69 Livy 22.5, 27.48; Caes. *Bell Afr.* 15; Dion. Hal. *Ant. Rom.* 9.50.3.
70 Livy 3.7.3, 3.54.10, 5.37.4.
71 Veg. *Mil.* 2.7.
72 Cf. the hard fighting the standard bearers successfully endured at Joseph. *BJ* 6.225.
73 Siebold (2007), 287.
74 Siebold (2007), 288.
75 Siebold (2007), 289.
76 King (2006), 500.
77 Livy 5.55.1.
78 Webster (1969): 135.
79 Hand gestures may have also been used, though the evidence is wanting, e.g., Ammianus 24.6.13; Lee (1996): 201.
80 Caes. *BG* 7.87; App. *B Civ.* 7.80.
81 App. *B Civ.* 2.78, as Caesar's forces were ordered to charge, 'the standard-bearers and officers put themselves in motion and exhorted their men'.
82 Caes. *BG* 2.20; Veg. *Mil.* 2.7; c.f. also Lewis and Short (1956), s.v. *signum.* It is interesting to note that the same Latin word is used for both the standard and the watchword.
83 Livy 27.46.
84 App. *B Civ.* 2.76.
85 Caes. *BG* 7.52.
86 Caes. *BG* 2.26, 7.49, 7.52, 7.81, 7.83; *BCiv.* 3.89; App. *B Civ.* 2.76; *Syr.* 31; Livy 37.41.

87 App. *Syr.* 33; *B Civ.* 2.83.
88 Caes. *BG* 4.13, 7.45, 7.52.
89 Caes. *BG* 2.26.
90 Caes. *BCiv.* 1.45, 3.93–95; App. *B Civ.* 2.81; Plut. *Pomp.* 69; cf. also Plut. *Pyrrh.* 16.8–10.
91 Caes. *BG* 2.22.
92 see n. 47.

Bibliography

Anderson, J. K. 1970: *Military Theory and Practice in the Age of Xenophon*, Berkeley.

Bell, M. J. V. 1965: 'Tactical reform in the Roman republican army', *Historia* 14: 404–422.

Brand, C. E. 1968: *Roman Military Law*, Austin.

Breeze, D. J. 1969: 'The organization of the Legion: The first cohort and the equites Legionis', *JRS* 59: 50–55.

Connolly, P. 1998: *Greece and Rome at War*, London.

Cornell, T. J. 1995: *The Beginnings of Rome: Italy and Rome from the Bronze Age to the Punic wars (c.1000–264 BC)*, London.

Daly, G. 2002: *Cannae: The Experience of Battle in the Second Punic War*, London.

Domaszewski, A. von. 1967: *Die Rangordnung des römischen Heeres*, 2nd edn. by B. Dobson, Köln.

Donaldson, G. H. 1988: 'Signalling communications and the Roman imperial army', *Britannia* 19: 349–356.

Edwards, H. J. trans. 1917: *Caesar. The Gallic War* (*Loeb Classical Library* 72), Cambridge, MA.

Goldsworthy, A. K. 1996: *The Roman Army at War 100 BC–200AD*, Oxford.

Hammond, N. G. L. and H. H. Scullard (eds.) 1970: *The Oxford Classical Dictionary*, 2nd edn., Oxford.

Haynes, I. P. 1993: 'The Romanisation of religion in the Roman imperial army from Augustus to Septimus severus', *Britannia* 24: 141–157.

Kagan, K. (2006). *The Eye of Command.* Ann Arbor.

King, A. 2006: 'The word of command: Communication and cohesion in the military', *Armed Forces & Society* 32(4): 493–512.

Koon, S. 2011: 'Phalanx and Legion: The "Face" of Punic War Battle'. In B. D. Hoyos (ed.), *A Companion to the Punic Wars*, Oxford and Malden MA: 77–94.

Krentz, P. 1991: 'The *Salpinx* in Greek Warfare'. In V. D. Hanson (ed.), *Hoplites: The Classical Greek Battle Experience*, London: 110–120.

Lee, A. D. 1996: 'Morale and the Roman Experience of Battle'. In A. B. Lloyd (ed.), *Battle in Antiquity*, London: 199–218.

Lewis, C. T. and C. Short, 1956: *Latin Dictionary*, Oxford.

Marshall, S. L. A. 1968: *Men against Fire*, New York.

Mathiesen, T. J. 1983: *Aristides Quintilianus. On Music. Translation with Introduction, Commentary and Annotations*, New Haven and London.

Myers, R. 2006: 'Deepeeka Cornu', *Roman Army Talk* online discussion list, 12 April 2006. Available at: http://www.romanarmy.com/rat/viewtopic.php?t=8114 [Accessed: 21 May 2007].

Nicholson, H. J. 1997: *Chronicle of the Third Crusade, A Translation of the* Itinerarium Peregrinorum et Gesta Regis Ricardi, Ashgate.

Oakley, S. P. 1998: *A Commentary on Livy Books VI–X. Volume II: Books VII–VIII*, Oxford.
Ross, C. D. 2001: *Civil War: Acoustic Shadows*, Shippensburg, PA.
Rostovtzeff, M. 1942: 'Vexillum and victory', *JRS* 32: 92–106.
Sabin, P. 2000: 'The face of Roman battle'. *Journal of Roman Studies*, 90: 1–17.
Siebold, G. L. 2007: 'The Essence of Military Group Cohesion'. *Armed Forces & Society* 33(2): 286–295.
Smith, W. *et al.* 1891: *A Dictionary of Greek and Roman Antiquities*, London.
Speidel, M. P. 1976: 'Eagle bearer and trumpeter', *Bonner Jahrbücher* 176: 123–163.
Taylor, M. J. 2014: 'Roman infantry tactics in the mid-republic: A reassessment', *Historia* 63(3): 301–322.
Van Epps, G. 2008: 'Relooking unit cohesion: A sensemaking approach', *Military Review* (November–December 2008): 102–110.
Webster, G. 1969: *The Roman Imperial Army*, London.
Weick, K. E. 1993: 'The collapse of sensemaking in organizations: The Mann Gulch Disaster', *Administrative Science Quarterly* 38: 628–652.
Zhmodikov, A. (2000). 'Roman Republican heavy infantrymen in battle (IV-II centuries BC)', *Historia* 49(1): 67–79.

7 The Legionary Standards as a Means of Religious Cohesion

Ben Greet

An important part of unit cohesion in any military organisation is social cohesion between the individual members of that unit.[1] MacCoun defines social cohesion as:

> the nature and quality of the emotional bonds of friendship, liking, caring, and closeness among group members. A group is socially cohesive to the extent that its members like each other, prefer to spend their social time together, enjoy each other's company, and feel emotionally close to one another.[2]

Religion can have an important impact on this aspect of unit cohesion, as the sharing of religious belief provides a common bond through which these forces bonds of friendship can form, leading to a greater cohesion within the unit. However, when applied to the Roman legion, the idea of religious cohesion becomes problematic, due to the polytheistic nature of Roman society. If the individual legionaries each have differing religious beliefs and ideals, a barrier is created that would disrupt the easy forming of religious and social cohesion within the unit.

Most modern studies on the Roman army emphasise the importance of the standards, and the cult surrounding them, in fostering religious cohesion within the legions through their belief and participation in this cultic activity.[3] Most of these stem from the quintessential studies on the army by Webster and Watson in the 1960s[4] or Helgeland's in-depth study of the religious role of the standards in the legions in 1978.[5] However, while studies like Helgeland's have emphasised the standards' religious nature as the reason for their importance in the creation and maintenance of religious cohesion in the legion, none of them precisely define the standards' religious role, or discuss the possible larger ramifications of their religious nature.[6]

A possible interpretation of an intaglio gem may provide a glimpse into the conceivable extent that the standards could foster religious cohesion within the legions. The gem shows the image of a Menorah, with three of its candlesticks replaced by what appear to be military standards.[7] Since it has been shown that Jews were part of the legions since the time of Tiberius, with some

DOI: 10.4324/9781315171753-8

even serving in Sardinia, this gem may have belonged to a legionary with Jewish beliefs.[8] If so, how did he reconcile integrating the standards, essentially a 'pagan' object of worship, with his personal cultic item of the Menorah? In this study, I intend to show that the standards, rather than just creating a separate military cult, performed a much broader religious role and represented a group of deities that fostered inclusiveness of religious belief. Through this inclusiveness, they built religious cohesion, which in turn built social cohesion, leading to a stronger overall cohesion within the legion.

Before beginning my analysis, it must be pointed out that I will be focusing mostly on the *aquila* standard and other legionary standards,[9] rather than the smaller unit standards. This is partly because more evidence exists for the religious nature of the legionary standards, but we can also presume that what religious beliefs surrounded the standards of the entire legion filtered down to the standards of individual units. Additionally, considering the numerous differences in both the organisation and culture of the manipular and imperial legions, this study will concentrate purely on the imperial legions (mostly post-Augustus). This is again due to the nature of the evidence, which is far more numerous for the post-Marian reform legions.[10]

Most of the studies that discuss the religious nature of the standards focus on two passages, one from Tertullian and one from Tacitus. The passage from Tertullians's *Ad Nationes* states that the standards are the religion of the legion and that the legionaries 'prefer them to Jupiter himself'.[11] The passage from Tacitus' *Annals* states that eagles, and thus the *aquila* standard they appear on, are *numina legionis*, which is usually translated as 'deities of the legions'.[12] These passages, and the appearance of the standards as *genii* (divine spirits tied to individual objects, people, or concepts) on some inscriptions,[13] seems to be enough for some to establish that the cult of the standards existed, and that this cult added to religious cohesion in the legions. While this traditional interpretation is adequate, it would merely provide an adjunct to each legionary's (usually) polytheistic worldview, rather than replacing the most important deity or deities in their individual religious beliefs. However, a closer reading of the evidence surrounding the religious nature of the standards points towards a different interpretation of their function and position in Roman religious thought, one that hints at their contributing to the cohesion of the legions.

In order to determine precisely how the standards were viewed in Roman religious thought, we must first examine the religious terminology that surrounds them in the sources. The building in which the standards are housed is named in a second century AD inscription from Reculver as the *aedes principiorum*.[14] It is the building that is created around the standards when they are struck in place during a campaign, and the central space within the *principia* of a legionary fort.[15] It describes the building around which the religious life of the legion revolved, with every sacrifice and festival taking place outside the *aedes* in full view of the *aquila* and standards.[16] A brief survey of Roman forts, from Roman Britain to Dura-Europos in

Syria, shows that the *aedes* has a standardised form in (nearly) every legionary fort or camp.[17] At the back of the *principia* were three to five rooms with the central room as the *aedes principiorum*.[18] This uniformity of planning and structures implies a uniformity of ritual involving the standards. The standards were kept in the *aedes* and taken out when any religious ritual was taking place within the camp.[19] Also in the second century AD, we can see visual representations of these rituals taking place on Trajan's column[20] and from a smaller garrison on the Antonine Wall.[21]

The fact that the Reculver inscription refers to the building housing the standards as an *aedes* is a clue to the position of the standards in Roman thought. The *aedes* was the 'temple-building', meaning the structure that housed the statue of the deity.[22] Rituals are then conducted in front of the *aedes*, without opening or going within it.[23] This seems to fit with most of the practices surrounding the standards and would imply, by the use of *aedes*, that the standards are synonymous with the cultic objects within other *aedes*, thus they are the equivalent to the statue of a deity. This is corroborated by a passage in Cicero, where he mentions that Cataline stored the *aquila* in his home, within his personal '*sacrarium*'.[24] Dyck's commentary points out that that the '*sacrarium*' is a room which stores sacred objects, implying that the standard itself is a cultic object.[25] This also matches with a passage of Dionysius, who says the standards are 'like a statue of the gods' ('*ιδρυματα θεῶν*').[26] Additionally, Wheeler believes that the phrase *infesta signa*, seen in numerous sources,[27] attests to the standards (*signa*) as 'indicators of the visible presence of the gods'.[28] However, it may be that the building was referred to as an *aedes* because it housed actual cultic statues. Certain forts have produced various other cultic objects from within the *aedes*, such as a figure of Jupiter from Murrhardt, one of a *genius* from Kapersburg, and some part of a Hercules sculpture from Kögen.[29] Thus, the term *aedes* may have been used to describe these buildings due to these statues rather than the standards.[30] Even if the word is referring to the standards, they still do not perfectly fit the rituals usually associated with cultic objects, which were never removed from the *aedes*,[31] whilst the standards were portable and free to move anywhere.

A passage from Tacitus might enlighten us on whether the standards were cultic objects or not.[32] During his description of the Germanic Wars he relates the story of an envoy of the Roman Senate being set upon by mutinous legionaries. The ambassador, to shield himself, runs into the *aedes principiorum* and grabs hold of the standards in order to 'protect himself under their sanctity (*religione*)'. His plan succeeds, but Tacitus posits that if it had not then 'the blood of an envoy of the Roman people … would … have stained the *altaria deum*'.[33] Since the ambassador had clasped himself around the standards, Tacitus must have been referring to the standards as the '*altaria*'. Although sometimes translated as 'altar', the word *altaria* is more complex. It is not merely a synonym for *ara* and both Pliny and Tacitus distinguish between *ara* and *altaria*.[34] Instead, its uses vary with some common features apparent in each instance of its use: (1) it is often used to distinguish one altar from others;[35] (2) an element of oath taking is often included;[36] and, lastly, (3) there seems to be an implication of portability in various instances.[37] With the passage in Tacitus, there seems to be two possibilities for why he has chosen *altaria*. The first is that he may be trying to distinguish the standards as the 'high altar' of the legions, since it is in front of them that the legionaries take their oath. However, he may also be trying to imply portability. The standards, since they are portable, differ to a regular *ara*, which appear to be inherently static.[38] This portability may even relate specifically to the eagle atop the *aquila* standard. The image of an eagle on an altar between two standards seems to have been a military symbol, appearing on numerous rings and on Vespasianic and later coinage.[39] This may have reflected the physical reality of the eagle being detached from the *aquila* and placed on an altar within the *aedes principiorum*. This detachability of the eagle from the *aquila* standard is demonstrated when Catiline removes it and places it in his house.[40] This may also explain why the relief of an eagle is found in the *principia* at the fort in Corbridge (replacing the actual eagle-standard of the legion, housed with the first cohort).[41] It may be, though, that Tacitus was using *altaria* in all these forms, thus identifying the standards as portable high altars for the legions. This interpretation of the standards as *altaria* also fits with the Dionysius passage mentioned above, as the Greek noun *ἱδρυμα* can be translated as shrine rather than statue.[42]

The visual depictions of military sacrifices complicate matters. The most prominent example is the depiction from the Antonine Wall in Scotland.[43] It shows the leader sacrificing to an altar that looks no different to a standard Roman *ara* – not *to* or *on* the actual standards, as the previous discussion of the use of *altaria* in Tacitus would imply. These altars are burning, like some descriptions of the *altaria*,[44] and it may be that Tacitus is instead referring to these altars as the *altaria deum*. However, since they look so much like the static Roman *ara*, which always remained outside of the *aedes*, why would Tacitus state that they were inside of the *aedes principiorum* with the standards, while also indicating that the ambassador's blood would fall upon it if he was clinging to the standards themselves? Instead, what seems more likely is that while the standards were *altaria*, a high-altar that could be moved, whilst an actual *ara* was used for sacrifices. In fact, a passage from Tacitus states that after a speech by Vitellius in a legionary camp a sacrificial bull 'escaped from the altars (*altaribus*)',[45] which seems to imply an area of altars rather than one, perhaps the area by the standards.

Despite this, the original criticism provided by the visual evidence remains, as the standards still do not fit the normal form of an *altaria*, because no rituals were actually performed *on* them. However, they also do not fit with the standard idea of cultic objects either (if that is what the word *aedes* and Dionysius are referring to), as they can be removed from the *aedes* and brought to the *ara*. Instead, they seem to be unique in their religious nature:

a combination of both *altaria* and cultic statue. Additionally, increasing their uniqueness, Tacitus refers to them as an altar 'of all the gods' and while some altars are dedicated to multiple deities, they never seem to be dedicated to all at once.[46]

So, if the standards are a combination of an *altaria* and a cultic statue, with the ability to create a sacred space when placed, to which deity does this space belong, and which deity do they represent? For many, the obvious answer would be Jupiter Optimus Maximus. Much of the scholarship refers to Jupiter as the main deity of the army[47] and some scholars assume that the eagle on the *aquila* represents Jupiter.[48] However, there are significant problems with making these assumptions. Although the eagle appears with Jupiter more than other deities,[49] he is not the only deity that can claim a connection to the bird. Many others, including Apollo and Mars, are shown with or using the eagle in literary or visual depictions.[50] Additionally, not even the thunderbolt in the eagle's claws on the standard is enough to definitively prove a connection to Jupiter. In Roman religious thought, a number of deities had the power to use the thunderbolt.[51] Apollo even had access to both the thunderbolt and the eagle.[52] Additionally, the idea that Jupiter Optimus Maximus was the premier deity of the legions is too simplistic a version of Roman army religion. We know from the *Feriale Duranum*, a religious calendar of a detachment of soldiers in Dura-Europos from the third century AD,[53] that a large number of deities were worshipped in a military context, presumably with the standardised ceremony discussed above.[54] We also know from other evidence, like the cultic statue of Heracles found in the *principia* mentioned above,[55] or the worship of Elagabal in Pannonia,[56] that the legionaries worshipped other deities not mentioned on the calendar. However, as far as we can tell, the rituals surrounding these festivals were the same, with the standards being taken out of the *aedes* and into the courtyard and sacrifices made at the *ara*.[57] Even if the *ara* itself was dedicated to an individual deity, as we know many were dedicated to Jupiter,[58] the standards were malleable enough to be used as divine representation, presumably, of all the deities worshipped by the legions. This also fits with Tacitus' assertion that they were *altaria* of (presumably) all the gods.[59]

What, then, gives the standards this unique position in Roman theological thought as a part-*altaria* and part-cultic object that can represent any deity? It is possible that the answer lies in the *genius* of the standards, seen on a number of inscriptions from the second century AD.[60] We know that the standard possessed some inherent divine qualities, as in certain omens they act of their own accord, specifically the stories of Flaminius,[61] where they refuse to be pulled from the ground, or of Crassus at Carrhae, where again they refused to be lifted and then turned about of their own volition.[62] In Apuleius' philosophy the cause of prodigies and the movement of physical objects are the result of intermediary spirits (*daimones* that he equates to *genii*).[63] Thus, the *genius* may be what allows the standards to turnabout of their own accord or to not be moved if they do not wish it. Thus, it may be this *genius*, acting as an intermediary spirit with other deities, that creates the unique position of the standards in Roman religious thought.

However, there are number of problems with this interpretation. One is that the usual representations of the *genii* were anthropomorphic,[64] whereas the standards are never depicted in an anthropomorphised form, but only as direct representations as standards. The most significant problem, though, comes from the fact that the *genius* is not a unique quality in Roman religion. Every place,[65] every person,[66] and every object, had its own personal *genius*.[67] Thus, why would the standard's *genius* provide any unique qualities that other *genii* do not? Even if they are the method by which the legions communicate with higher divine beings, it is certainly not the only method of communication. If it was, each legionary inscription to the gods would most likely make reference to the *genius*. While this is seen on some inscriptions,[68] it is certainly not present on all.[69]

If it is not their *genius*, then, that provides the standards with their unique position, it must be attributed to something else. Returning to the previously mentioned and often quoted Tacitus passage might shed light on their exact religious nature. In it, he reports an omen, again in the German Wars, where Germanicus sees eight eagles fly in the direction that his eight legions must attack. He states that the legions must follow them as they are the '*propria numina legionum*'.[70] As previously mentioned, this passage is used in nearly every examination of the cult of the standards and is most often translated as 'true deities of our legions'.[71] This is a simple enough idea and would indicate that eagles, and thus the *aquila* (and possibly the other standards) were deities themselves, worshipped alongside other legionary deities at the sacrifices and rituals described above. This interpretation would also explain the Tertullian passage.[72] Tacitus may even be referring to the *genii* just discussed. However, the word *numen* is a more complicated concept than this translation allows.[73] One definite feature that seems common in all interpretations is that it is not a purely divine concept. Seemingly in every mention of the word, especially when it appears in Tacitus,[74] it always has some sort of physical connection with the mortal plane.[75] The best example of this being the *numen* within a living emperor, he is not a god but has divine qualities within his mortal form.[76] Therefore, it seems to be a manifestation of the divine on the physical plane, hence why it is attached only to physical objects (like the eagles in this omen, statues of gods, and the emperor himself).

There are, then, two probable interpretations of the *numina legionum* in this passage. The first would be as 'the divine will/action/power of the legions'. This idea of the *numen* as the 'divine will of a deity' is the standard definition given by Scheid[77] and this seems reflected in Tacitus' use of the word throughout both the *Annals* and the *Histories*.[78] His *numen* mostly seems to be the physical manifestation of a deity's will or power. Since he used *legionum*, he must mean the *numen* 'of the legions'. Considering there are eight eagles and eight legions, it would be reasonable to assume he was referring to the divine representation of the legions in the form of the *genius.*[79] Thus, the eagles are the *numen*, or physical manifestation of the divine power of the *genius* of the legions on the mortal plane. However, this particular use of *numen* seems connected more to mortal beings that possess a divine power that allows them to become a god after death, like the emperor Augustus.[80] Since the *genii* were mostly divine, it would seem strange to use *numen* in connection with them in this way. The second, and more likely, interpretation is that *numen* was not being used as a direct synonym for a deity, as in 'deities of the legions', but to refer to a group of deities that centre around something in the physical plane. There are examples of the word being used similarly in different contexts, such as references to *numini Augusti*,[81] *numen Idaeum*,[82] and the especially relevant *numina castrorum*.[83] In this interpretation, and as Fishwick points out with the *numini Augusti*,[84] it would be referring to a group of deities that protect the legion as a whole, the *Di Legionum*, or a group we see actually referred in inscriptions, the *Di Militares*.[85]

The eagles, and thus the *aquila* as well (and by proxy the standards that accompany it), may, then, be the physical representation of this abstract group of deities connected to the legions: the *Di Legionum*. This interpretation explains some of the religious aspects of the standards explored above. Just as Jupiter has a statue representing him inside his *aedes*, the *Di Militares*, as a whole, are represented by the eagle in the *aedes principiorum*. Hence, Tacitus called them an *altaria* to all the (presumably relevant) military gods and referred to them either as representative of those deities' divine will; their physical representation, like a statue; or as representative of the actual deities in the form of an eagle, their usual mode of representation on earth. In fact, on an inscription of the second century AD, the *Di Militares* are included with the *genius* of the eagle and *sancti signa* ('sacred standards').[86] In another passage of Tacitus, Antoninus Primus prays through the *signa* to the *belli di* ('deities of war'),[87] what seems like a synonym for the *Di Militares*. Thus, essentially, the eagle and standards are the cultic statue and representative of all the possible deities that could be ascribed as one of the *Di Militares*, giving this abstract group of deities a tangible focal point within the religion of the Roman army.

So how does this alternative interpretation of the religious nature of the standards impact on the religious cohesion of the legions as a whole? The basic nature of polytheism creates an individualistic experience when it comes to religion. In practice, this meant that a legionary from a Syrian background would worship different deities, with different rituals, than a legionary from Gaul or Italy. In turn, those two legionaries from Gaul and Italy would also worship slightly different deities. Additionally, even those from a shared cultural background might emphasise one particular deity over another or belong to a particular cult that others did not. As mentioned previously, this differing religious belief creates a barrier during social interaction, which, in turn, lowers social cohesion within the unit.

This is where this interpretation of the religious nature of the standards becomes important. If the eagle and standards represented *all* of the possible deities that could be labelled '*Di Legionum*' or '*Di Militares*' then each individual's personal religious proclivities could be placed under that abstract banner. Whichever deity they believed watched over them during warfare was represented by the eagle and the standards. Additionally, since the standards seem to have been used in every religious festival or ritual, their personal deities would be present even if the festival was not dedicated to them. This creates a powerful method of religious, and therefore social, cohesion amongst the legionary community. Rather than the inherent nature of polytheism dividing the community through religious disharmony, the standards create a method of binding these different religious beliefs into a cohesive whole, focused on the legionary community. In this way they are working to build social cohesion on two levels: (1) they reduce the amount of friction that is inherent within a community with differing religious beliefs or priorities by providing an ambiguous focal point of religious worship that allows for a broad spectrum of belief; and (2) they build religious, and therefore social, cohesion by binding together all of these differing religious beliefs into a central concept of the '*Di Legionum*' or '*Di Militares*', which allows the legionary community to participate in communal religious experiences, despite their differing belief systems. Both these methods facilitate the building of the strong emotional bonds that MacCoun believes are necessary for the social cohesion of a unit.

Perhaps the best and most extreme example of this process would be the (possibly) Jewish intaglio gem mentioned at the beginning of this chapter. Although monotheistic, the individual who owned the gem felt he could include the military standards within his own personal religious beliefs, even though the polytheistic nature of legionary religion was at odds with his beliefs. In this case, the religiosity of the standards, and the importance they held in both the legionary community and for the communal religious experience of the legionaries, allowed him to include what was inherently a pagan symbol within his personal religious experience, despite its supposed incompatibility with the wider tenets of his religion.

Notes

1 MacCoun, Kier, and Belkin (2006).
2 MacCoun (1993): 291.
3 Goldsworthy (1996): 255–6; Gilliver (2007): 187; Hekster (2007): 352; Stoll (2007): 457–8.
4 Watson (1969): 128–31; Webster (1969): 133–8.
5 Helgeland (1978).
6 Helgeland (1978); Le Bohec (1989): 246.
7 Henig (1983): fig. A.
8 While it may have been owned by a Roman demonstrating victory over the Jews, like the Menorah on the Arch of Tiberius, I think the incorporation of the standards into the Menorah seem to indicate a melding of beliefs rather than a defeat of Jewish ideas. Judaism has been shown to be more flexible than previously assumed, Schwartz (2001); Levine (2003); Schoenfeld (2006): 117. While

Tiberius forcibly conscripted 4,000 Jews in AD 19, Joseph. *AJ.* 18.84, at least one inscription indicates voluntary service, *CIL* 16; Schoenfeld (2006): 117–18; 120–2. A Jew from Syrian served in the *Legio I Adiutrix* under Nero, *CIL* 16.8; and other inscriptions do the same, *CIL* 2.920. Joseph. *AJ.* 17.2 refers to Jewish military colonists. Additionally, the Jews may have already been familiar with the concept of a cultic standard, since Josephus uses the same word (*σημαία*) as Lucian when describing the Roman military standards, Joseph. *AJ.* 18.55; *BJ.* 2.169; Lucian, *Syr. D.* 33; 49.

9 This is except for the *imagines*, which are directly connected with the imperial cult.

10 There is also a distinct difference between the standards of the manipular legions and the imperial legions, as noted by Pliny in his description of the five animal standards of the pre-Marian legions (Plin. *HN.* 10.5). In fact, since these standards are only mentioned by Pliny (who was two hundred years removed from the Marian reforms) and do not fit perfectly with Polybius' description of the manipular legions, there is significant reason to doubt even the existence of these five animal standards of the pre-Marian legions.

11 Tert. *Ad Nat.* 1.12; *Apol.* 16; Webster (1969): 133–4; Helgeland (1978): 1477.

12 Tac. *Ann.* 2.17; Watson (1969): 128; Helgeland (1978): 1477. Henig (1984): 90 rejects that the standards were ever 'gods', but still emphasises their religiosity.

13 *CIL* 7.103; 1031; 2.6183; Helgeland (1978): 1477.

14 Wright (1961): 191, no.1; Birley (1978): 1476; 1539. An Egyptian legionary papyrus refers to it as the *aedem aquilae*, notably it is connected directly to the standards, Wheeler (2009): 260. In modern literature, the building is referred to as the *sacellum;* however, this is not mentioned in any of the ancient sources.

15 Watson (1969): 131; Webster (1969): 133–4. Caesar's *signa tollere* and Polybius' description point to this interpretation; Caes. *BCiv.* 2.20; Polyb. 6.2.

16 Herodian, 4.4.5; Joseph. *BJ.* 6.6, the organisation of the camp seems not to have changed in Josephus' period; thus, this can be presumed to have existed in the pre-Augustan period.

17 Radford (1936): 11; Birley (1936): 16; (1960): 22; Richmond (1953): 5–6; Robertson (1964): 90, 118; Helgeland (1978): 1491; Dore and Gillam (1979); Johnson (1983) shows this uniformity across Britain and Gaul.

18 Johnson (1983): 111.

19 Beard, North, and Price (1998a): 326–7. This matches the ritual practice of the Near Eastern cultic standards, the *semeia*, Lucian, *Syr. D.* 33; Dirven (2005): 120.

20 Beard, North, and Price (1998a): 327.

21 Henig (1984): 86–7.

22 Beard, North, and Price (1998b): 78.

23 Barton (1989): 68; Beard, North, and Price (1998b): 78. This was because the altar was usually housed outside of the *aedes*. Additionally, although a *templum* can exist without an *aedes*, an *aedes* cannot exist without a *templum*, Barton (1989): 67–8. In this case, it seems that the camp itself is the *templum*, as it was sacred space, Beard, North, and Price (1998b): 86). Thus, when the camp was mobile it was again the standards that imparted this sacredness to the space, which then became a *templum*.

24 Cic. *Cat.* 1.24; 2.6.13.

25 Dyck (2008): 110.

26 Dion. Hal. *Ant. Rom.* 6.45.2; Wheeler (2009): 256.

27 Caes. *BGall.* 3.93; 6.8.6; 7.51.3; Cic. *Font.* 16; Livy 2.30.11; 26.13.11; Luc. 3.330; Flor. 1.17.2; Sall. *Cat.* 60.2; Veg. 3.20.1.

28 Wheeler (2009): 255.

29 Johnson (1983): 113.

30 Stoll (2001): 262 thinks all the terminology for the sanctuary of the standards in literary sources is unreliable, but he provides no viable alternative, and thus we must deal with the evidence we have; Wheeler (2009): 260.
31 Beard, North, and Price (1998b): 78.
32 For notes on Tacitus' military service and authorial intention that relate to this passage, see Tac. *Agr.* 45.5; *Ann.* 11.11; Birley (2000): 234–5.
33 Tac. *Ann.* 1.39.
34 Livy 10.38.9–11; Plin. *HN.* 10.17; Tac. *Ann.* 16.31.
35 In Livy 10.38.9–11, the Samnites have a grove with many *arae*, but the soldiers are led to the *altaria* to swear their oath. Juv. 8.156, seems to distinguish Jupiter *altaria* from other *ara*. In Tac. *Ann.* 16.31, Servilia places her hand on both the *altaria* and the *ara* to swear an oath. Griffiths (1975): 196, seems to think Apul. *Met.* 11.10 is also using *altaria* to denote a 'high altar'.
36 Livy 10.38.9–11; 21.1.4; Tac. *Ann.* 11.9; 16.31.
37 Apul. *Met.* 11.10; Curt. 3.3.9; 5.1.20; Tac. *Hist.* 2.70.
38 Höcker and Prayon (2002): 544.
39 A few prime examples are: *RIC* Vespasian 847; *RPC* Vespasian 1973; Henig (1984): no.705. It is from the iconography of the gems, showing an eagle on an altar between two standards, which gives rise to the possibility that it is a military symbol, this makes sense since it is first adopted by Vespasian, who had a close relationship with the legions and these gems were found in legionary forts.
40 Cic. *Cat.* 1.9.24; 2.6.13.
41 Richmond (1943): 153–4. The eagle relief in York Museum may have also some from the Roman fort.
42 See above, n. 26. *LSJ* s.v. *ιδρυμα*. The noun seems to indicate a statue in Aesch. *Per.* 811, but a shrine in Aesch. *Ag.* 339; *Lib.* 1036; Eur. *Bac.* 951; Hdt. 8.144. Plato *Leg.* 717b, also uses it to refer to a private shrine, rather than a public one. The word seems just as interchangeable as the Latin ones discussed and, therefore, does not change our interpretation.
43 Henig (1984): 86–7; Beard, North, and Price (1998a): 327.
44 Curt. 5.1.20; Lucr. 4.1233–1239; 6.751; Petr. *Sat.* 135; Tac. *Hist.* 2.2–3; Verg. *Ecl.* 1.43; *OLD*: 106.
45 Tac. *Hist.* 3.56.
46 Paus. 1.34.3; Höcker and Prayon (2002): 544. Even within the legion, there is an altar to the *genius* of the emperor and the standards; Helgeland (1978): 1491.
47 Webster (1969): 135; Helgeland (1978): 1473; Fears (1981).
48 Webster (1969): 135; Helgeland (1978): 1473.
49 Ach. Tat. 2.12.2–3; 2.18.1; Apul. *Met.* 3.23; 6.15; Arr. *Anab.* 2.3.3–7; Ath. 13.566d; Cic. *Div.* 1.106; *Leg.* 1.1.2; Clem. *Recog.* 10.22; *Eleg. Maec.* 1.87; German. *Arat.* 315–320; 688; Lucian *Dial. D.* 6; *Icar.* 1; 22; *Sac.* 5; Manilius, 1.343; 5.486–503; Mart. *Ep.* 10.19; Ov. *Am.* 1.10; *Ars Am.* 3.419–422; *Met.* 6.511–518; Paus. 3.17.4; 5.11.1; 5.22.5; 5.22.7; 8.30.2; 8.31.4; 8.38.7; Plin. *HN.* 2.56; 10.4; 6; Stat. *Theb.* 3.532; Val. Flacc. 1.149–164; 2.409–417; Verg. *Aen.* 1.28; 394; 5.255; 9.564; 12.238–255; Bailey (1980): nos.1026–1028; 1056; 1210; 1224–5; 1229; 1232; 1250–1; 1523; BMC 6; 9; 19; 26; 40–41; 78; 165; *CNG* 67; 126; 169; 1109; *LIMC* Ammon 143; Castores 79; Dodekatheoi 23; Iuppiter 36–37; 53–55; 74–75; 77; 99; 105–107; 117a–117b; 132; 166; 168; 173; 187; 196; 204; 216; 220; 234; 236; 246a; 278; 359; 409; 412–413; 482; 504; 526; Minerva 283; Sarapis 104; Zeus 292; 351; 355; 393; 400; Zeus (in *peripheria orientali*) 141; 212; *RIC*, Anon. II.5; Hadrian 495; 497; Antoninus 711; Commodus 187; 525; Pescennius 41; Septimius 196, to name a few.
50 Hyg. *Fab.* 10; Stat. *Silv.* 2.5.178–180; BMC 1; 4; 21; 34–37; *LIMC* Arkadia 1; Elegabalos 2; 3a; Palmyra 3; *RIC* Clodius 23c; Hadrian 284a; Vespasian 6; *RPC* I, 2907; 1808–1809; 1811; 1813; 1815; *RPC* II, 1777.

51 Aesch. *Eum.* 827–8; Apollod. *Bibl.* 1.9.26; 3.10.4; Diod. Sic. 4.71.3; Eur. *Tro.* 77–81; 92–94; Hom. *Il.* 12.17–33; 14.385–386; *Od.* 4.506–507; 5.291–292; Just. *Epit.* 2.12; Mart. Cap. 1.7; 1.67; Paus. 10.23.1; Plin. *HN.* 2.138–139; Sen. *QNat.* 2.41; Serv. *Aen.* 1.42; 8.426; 11.45–46; Soph. *OT.* 464–470; Stat. *Theb.* 10.67–9; Ver. *Aen.* 1.42–5; McCartney (1932c): 214; Hekster and Rich (2006): 162. Diod. Sic. 11.14 even says that it could be used by any divine being.

52 Soph. *OT.* 496–470; Paus. 9.36.3; 10.23.1; Justin. 2.12.8; Prop. 4.6.29–30; Hekster and Rich (2006): 162. For the eagle and Apollo, see Aesch. *Pers.* 205–214; Diod. Sic. 16.27.2; Ov. *Met.* 2.452–524; *RRC*, 445/2; *SNG* Cop. 6; 837; Ravel (1947): no.49; Calciati (1983–1987): no.119; Zanker (1988): fig. 99.

53 Helgeland (1978) includes the text and translation. See Fink, Hoey, and Synder (1940); Nock (1952); Gilliam (1954); and Helgeland (1978) for dating.

54 The calendar includes festivals to the Capitoline Triad (never Jupiter alone), Dea Roma, Juno, Minerva, Mars, Neptune, Vesta, Salus, and a number of deified emperors. The calendar also includes a festival labelled the '*Rosaliae Signorum*' or 'rose-festival of the standards', which Hoey (1937) identifies as a spring festival.

55 Johnson (1983): 113.

56 *ILS* 4349; Beard, North, and Price (1998a): 327–8.

57 The surviving visual depictions seem to match, Beard, North, and Price (1998a): 327.

58 A number of altars dedicated to Jupiter were found in a fort in Britain, *RIB*: 815–817; 819; 822; 824–828; 830–831; 838–842; 843; Webster (1969): 277.

59 Tac. *Ann.* 1.39.

60 *CIL* 3.3526; 7591; 7.1031; Rostovtzeff (1943): 214; Nock (1952): 239–40; Helgeland (1978): 1477; 1491; Speidel (1978): 1547; *CIL* 2.6183 is dedicated to Jupiter, but in celebration of the *natalem aquilae*; Watson (1969): 130; Helgeland (1978): 1477.

61 Cic. *Div.* 1.77; and later in Livy 22.3.11; Val. Max. 1.6.6.

62 Cass. Dio. 40.17.3–18.2; Plut. *Vit. Crass.* 19.3; Val. Max. 1.6.11.

63 Apul. *De Deo Soc.* 133–137; 150–151. Interestingly, Aeschylus *Pers.* 811, uses the same word as Dionysus (*ιδρυμα*) to refer to statues of *δαίμοντες* (*daimones/genii*) and not *θεοί* (gods).

64 Speidel (1978): 1547.

65 The legionary camp, *CIL* 6.230; and other civilian buildings, *CIL* 2.2634; 3.1019.

66 Each emperor had his own *genius*, *CIL* 11.3303; Fishwick (2007): 249–50. Also, groups of people, like the soldiers themselves, *CIL* 2.5083; 3.1646; 15208; 6.227; 234; 7.103; 1031; *ILS* 2180; Le Bohec (1989): 245.

67 The *genii* of the legions, standard-bearers, cohorts, etc. have already been mentioned above, p.130-2.

68 *CIL* 7.203; Helgeland (1978): 1477.

69 As mentioned, *RIB*, 815–817; 819; 822; 824–828; 830–831; 838–842; 843 are all just dedicated to Jupiter Optimus Maximus; Webster (1969): 277.

70 Tac. *Ann.* 2.17.

71 Nock (1952): 239–40; Watson (1969): 128; Speidel (1976): 142–3; Helgeland (1978): 1474. However, Henig (1984): 90 goes against this opinion that they were actual deities.

72 Tert. *Ad Nat.* 1.12, see above n. 11.

73 The word has been much discussed, Warde-Fowler (1911): 118; Rose (1948): 13; Weinstock (1949); Dumézil (1966); Scheid (2003): 153.

74 Tac. *Ann.* 1.10; 1.73; 3.71; 4.37; 14.14 and 15.45, refer to statues of the gods; 13.41; 13.57; and 15.34, divine miracles; 15.36 and 16.25, divine inspiration; 15.74, divine revelation; 16.16, the gods' wrath against Rome; *Hist.* 2.61, Mariccus pretending at divine inspiration; 3.33, the Temple of Mephitis remaining standing; 4.57, divine vengeance enacted on the mortal plane; 4.65, the

mortal messenger of the divine; 4.84, the anger of a god as a physical threat and doing the bidding of a god in the physical plane; 5.5, thought about gods.

75 Other references to *numen* also seem to involve objects on the physical plane: poets can have *numen*, Ov. *Am.* 3.9.18; as well as springs, Ov. *Fast.* 5.674; *Her.* 15.150; *Met.* 1.545; virtuous men, Luc. 6.253–254; Rome, Luc. 1.199–200; the Roman people, Cic. *Red. Pop.* 18; the senate, Cic. *Phil.* 3.32; deceased loved ones, *CIL* 6.37965; *AE* (1976), 243; and trees, Stat. *Theb.* 6.93–94; 9.586–588.

76 Fishwick (2007); Hunt (2014).

77 Scheid (2003): 153; 157.

78 Tac. *Ann.* 1.10, 73; 3.71; 4.37; 13.41, 57; 14.14; 15.34, 36, 45, 74; 16.16, 25; *Hist.* 2.61; 3.33; 4.57, 65, 84.

79 Tacitus seems to be using *numen* as a synonym for *genius* when he speaks of Otho, Tac. *Hist.* 2.33; Ash (2007): 168.

80 Fishwick (1991): 387; 2007; Scheid (2003). Gradel (2002) thinks that the *numen* of the emperor was simply a synonym for the direct worship of the emperor as a god. While this seems mostly correct, he fails to account for the physical aspect of *numen*, i.e., in the examples he gives (Hor. *Carm.* 4.5.31–6; Plin. *Pan.* 52.6; Suet. *Cal.* 22.3) the emperor to whom the *numen* refers was alive at the time. Thus, the worship was directed at the divine power within a mortal man rather than a traditional god.

81 Abascal and Alföldy (2002): Fishwick (2007): 254. There is a military inscription which also refers to the *numinibus Augustorum*, *CIL* 7.103; Helgeland (1978): 1477.

82 Juv. 3.138; Fishwick (2007): 254.

83 *CIL* 13.6749; Fishwick (2007): 254.

84 Fishwick (2007): 251–4.

85 *AE* (1966): 355; Fishwick (2007): 254, like the Di Conservatores, *CIL* 8.10178; 17620; Fishwick (1991): 451; (2007): 254; or the Di Iuvantes, *CIL* 98.2226.

86 *CIL* 3.7591.

87 Tac. Hist. 3.10. Wheeler (2009): 257 does not perceive that Antonius turned towards the standards to pray and instead believes he was praying to both.

Bibliography

Abascal, J. M. and G. Alföldy, 2002: 'La Inscripción Del Arco'. In J. M. Abascal and G. Alföldy (eds.), *El Arco Romano de Medinaceli*, Madrid: 71–118.

Ash, R. 2007: *Tacitus, Histories: Book II*, Cambridge.

Barton, I. M. 1989: 'Religious Buildings'. In I. M. Barton (ed.), *Roman Public Buildings*, Exeter.

Beard, M., North, J., and S. Price, 1998a: *Religions of Rome, Volume 1: A History*, Cambridge.

Beard, M., North, J., and S. Price, 1998b. *Religions of Rome, Volume 2: A Sourcebook*, Cambridge.

Birley, A. R. 2000: 'The Life and Death of Cornelius Tacitus', *Historia* 49 (2): 230–247.

Birley, E. 1936: *Housesteads Roman Fort*, London.

Birley, E. 1960: *Chesters Roman Fort*, London.

Birley, E. 1978: 'The Religion of the Roman Army: 1895–1977', *ANRW* II 16 (2): 1506–1541.

Block, D. I. 1997: *The Book of Ezekiel: Chapters 1–24*, Grand Rapids.

Brownlee, W. H. 1986: *Ezekiel 1–19*, Waco.

Calciati, R. 1983-8: *Corpus Nummorum Siculorum: The Bronze Coinage* (3 vols.), Milan.

Christensen, D. L. 2002: *World Biblical Commentary, Volume VIB: Deuteronomy 21:10–34:12*, Nashville.

Davies, R. W. 1973: 'Minicius Iustus and a Roman Military Document from Egypt', *Aegyptus* 53: 75–92.

Dirven, L. 2005: 'ΣΗΜΗΙΟΝ, SMY', SIGNUM: A Note on the Romanization of the Semitic Cult Standard', *Parthica* 7: 119–136.

Dore, J. N. and J. P. Gillam, 1979: *The Roman Fort at South Shields: Excavations 1875–1975*, Newcastle upon Tyne.

Dozeman, T. B. 2009: *Commentary on Exodus*, Cambridge.

Dumézil, G. 1966: *Archaic Roman Religion, Volume 1*, Chicago.

Dyck, A. R. 2008: *Cicero: Catilinarians*, Cambridge.

Fears, J. 1981: 'The Cult of Jupiter and Roman Imperial Ideology', *ANRW* II 17 (1): 3–141.

Fink, R. O., Hoey, A. S., and W. F. Synder, 1940: 'The *Feriale Duranum*', *Yale Classical Studies* 7: 1–222.

Fishwick, D. 1991: *The Imperial Cult in the Latin West*, Leiden.

Fishwick, D. 2007: 'Numen Augustum', *Zeitschrift Für Papyrologie und Epigraphik* 160: 247–255.

Gilliam, J. F. 1954: 'The Roman Miltiary *Feriale*', *Harvard Theological Review* 47 (3): 183–196.

Gilliver, K. 2007: 'The Augustan Reform and the Structure of the Imperial Army', In P. Erdkamp, (ed.), *A Companion to the Roman Army*, Oxford: 183–200.

Goldsworthy, A. K. 1996. *The Roman Army at War 100 B.C. – A.D. 200*, Oxford.

Goodenough, E. R. 1958: *Jewish Symbols in the Greco-Roman Period, Volume VIII: Pagan Symbols in Judaism (The Second of Two Volumes)*, New York.

Gradel, I. 2002: *Emperor Worship and Roman Religion*, Oxford.

Griffiths, J. G. 1975: *The Isis-Book: Metamorphoses, Book XI*. Leiden.

Hekster, O. 2007: 'The Roman Army and Propaganda'. In P. Erdkamp (ed.), *A Companion to the Roman Army*, Oxford: 339–358.

Hekster, O. and J. Rich, 2006: 'Octavian and the Thunderbolt: The Temple of Apollo Palatinus and Roman Traditions of Temple Building', *CQ* 56(1): 149–168.

Henig, M. 1983: 'A Question of Standards', *Oxford Journal of Archaeology* 2 (1): 109–112.

Henig, M. 1984: *Religion in Roman Britain*, London.

Helgeland, J. 1978: 'Roman Army Religion', *ANRW* II 16 (2): 1470–1505.

Höcker, C. K. and F. T. Prayon, 2002: 'Altar'. In H. Cancik and H. Schneider (eds.), *Brill's New Pauly, Antiquity: Volume I, A-Ari*, Leiden: 543–549.

Hoey, A. S. 1937: *Rosaliae Signorum, Harvard Theological Review* 30 (1): 15–36.

Hunt, A. 2014: *Rethinking* Numen*: A Word for 'Thinking With'*, Classical Association Conference, Nottingham: 13–16 April.

Johnson, A. 1983: *Roman Forts of the 1st and 2nd Centuries A.D.*, London.

Joyce, P. M. 2007: *Ezekiel: A Commentary*, New York.

Le Bohec, Y. 1989: *The Imperial Roman Army*, London.

Levine, L. 2003: 'Contextualising Jewish Art: The Synagogues at Hammat-Tiberias and Sephorris'. In R. Kalmin and S. Schwartz (eds.), *Jewish Culture and Society under the Christian Roman Empire*, Leuven: 91–132.

MacCoun, R. J. 1993: 'What Is Known about Unit Cohesion and Military Performance', *Sexual Orientation and U.S. Military Personnel Policy: Options and Assessment*, Santa Monica: 283–331.

MacCoun, R. J., Kier, E., and A. Belkin, 2006: 'Does Social Cohesion Determine Motivation in Combat? An Old Question with an Old Answer', *Armed Forces & Society* 32 (4): 646–654.

McCartney, E. S. 1932a: 'Classical Weather Lore of Thunder and Lightning', *The Classical Weekly* 25 (23): 183–192.

McCartney, E. S. 1932b: 'Classical Weather Lore of Thunder and Lightning (Continued)', *The Classical Weekly* 25 (24): 200–208.

McCartney, E. S. 1932c: 'Classical Weather Lore of Thunder and Lightning (Concluded)', *The Classical Weekly* 25 (25): 212–216.

Nock, A. D. 1952: 'The Roman Army and the Religious Year', *Harvard Theological Review* 45 (4): 187–252.

Propp, W. H. C. 2006: *Exodus 19–40*, New York.

Radford, R. 1936: *Roman Site of Segontium*, London.

Ravel, O. E. 1947. *Descriptive Catalogue of the Collection of Tarentine Coins formed by M. P. Vlasto*, London.

Richmond, I. A. 1943: 'Roman Legionaries at Corbridge, Their Supply-Base, Temples, and Religious Cults', *Archaeologis Aeliana* 4 (21): 127–224.

Richmond, I. A. 1953: *The Roman Fort at South Shields*, Newcastle upon Tyne.

Robertson, A. S. 1964: *The Roman Fort at Castledykes*, London.

Rose, H. 1948: *Ancient Roman Religion*, London.

Rostovtzeff, M. I. 1943: *The Excavations at Dura-Europos*, New Haven.

Scheid, J. 2003: *An Introduction to Roman Religion*, Edinburgh.

Schoenfeld, A. J. 2006: 'Sons of Israel in Caesar's Service: Jewish Soldiers in the Roman Military', *Shofar: An Interdisciplinary Journal of Jewish Studies* 24 (3): 115–126.

Schwartz, S. 2001. *Imperial and Jewish Society, 200 BCE to 640 CE*, Princeton.

Speidel, M. P. 1976: 'Eagle-Bearer and Trumpter', *Bonner Jahrbucher* 176: 123–164.

Speidel, M. P. 1978: 'The Cult of the *Genii* in the Roman Army and a New Military Deity', *ANRW* II 16 (2): 1542–1555.

Stoll, O. 2001: *Zwischen Integration und Abgrenzung: Die Religion des Römischen Heeres im Nahen Osten*, St. Katharinen.

Stoll, O. 2007: 'The Religions of the Armies'. In P. Erdkamp (ed.), *A Companion to the Roman Army*, Oxford: 451–476.

Thompson, J. A. 1981: *The Book of Jeremiah*, Grand Rapids.

Warde-Fowler, W. 1911: *The Religious Experience of the Roman People*, London.

Watson, G. R. 1969: *The Roman Soldier*, Bristol.

Webster, G. 1969: *The Roman Imperial Army*, Norman.

Weinstock, S. 1949: 'Review of *Ancient Roman Religion* by H. J. Rose', *JRS* 39: 166–167.

Wheeler, E. L. 2009: 'Shock and Awe: Battles of the Gods in Roman Imperial Warfare, Part 1'. In Y. L. Bohec and C. Wolff (eds.), *L'Armée Romaine et al Religion sous le Haut-Empire Romain – Actes du Quatrième Congrès de Lyon organise les 26–28 Octobre 2006*, Paris: 225–267.

Wright, R. P. 1961: 'Roman Britain in 1960: I. Sites Explored: II. Inscriptions', *JRS* 51: 157–198.

Zanker, P. 1988: *The Power of Images in the Age of Augustus* (trans. A. Shapiro), Ann Arbor.

8 Looking for Unit Cohesion at the End of Antiquity[1]

Conor Whately

There has been a great deal of research on unit cohesion in classical and modern era armies. This is not the case for late antiquity. To address this gap, this chapter evaluates whether we have evidence for one form of unit cohesion in one specific time: primary-group unit cohesion, best represented in popular culture by the notion of militaries composed of 'bands of brothers', and in the sixth century CE, especially the age of Justinian. Our focus will include all land-based forces, whether the cavalry or the infantry. We are well informed about the sixth century, for which we have multiple histories including, especially, Procopius' *Wars*. These accounts, however, are not without their faults. Fortunately, we also have detailed military treatises, like Maurice's *Strategikon*, legal material, like Justinian's *Codex*, detailed excavation reports, like those published for el-Lejjūn in Jordan, and even some extensive collections of papyri, such as those uncovered at Nessana and Petra. A thorough investigation reveals that while we possess terminology for primary groups of eight to ten soldiers, it is not sufficient for deducing a role for primary-group horizontal bonding in unit cohesion. Rather, it underscores the need for further research on unit cohesion and military communities in late antiquity.

Introduction

In act 4, scene 3 of Shakespeare's *Henry the V*, there is a speech, the St. Crispin Day's speech, given by the eponymous king, in which we find the now famous lines:

> *And Crispin Crispian shall ne'er go by,*
>
> *From this day to the ending of the world,*
>
> *But we in it shall be remember'd;*
>
> *We few, we happy few, we band of brothers.*

The most famous part of that passage for military historians is the phrase, 'band of brothers', which, in the last few decades or so, has taken on a life of

DOI: 10.4324/9781315171753-9

its own, somewhat removed from its context. Stephen Ambrose used the phrase in the title to his 1992 World War II book, *Band of Brothers*, which was dramatized in an HBO miniseries co-produced by Steven Spielberg and Tom Hanks. In the play, the king, Henry V, gives the speech before the English troops at Agincourt immediately before the famous battle in 1415. The few in question are the men arrayed for England, who are few in comparison to their more numerous French foes. On the other hand, Shakespeare's 'few' are far greater in number (tens of thousands) than the 'few' usually adduced by military historians, who regularly use the phrase to refer to the bonding that is said to be characteristic of many or most militaries, and the prime motivating force in combat.[2] It is the later iterations of the phrase that are the impetus for this chapter, which is devoted to small-unit cohesion, of bands of brothers numbering eight to ten men.

It is bands of brothers, who form 'buddy' or 'peer groups' of common soldiers at the 'squad' level, that many identify as the key factor in success in combat, and which we will discuss here. The ties that bind soldiers is a theme found in countless war movies and accepted by the general public, as evidenced by the proliferation of movies like *Saving Private Ryan*, *Band of Brothers*, *Jarhead*, *Lone Survivor*, and *Hyena Road* that dramatize, in some capacity or other, this small unit cohesion. How well does such a model apply to late antiquity?

Siebold has proffered a standard model of cohesion with four components, each of which has an impact upon unit cohesion. To look at all four components would require a much longer and more detailed study. In the interests of space, the focus here with be on looking for evidence of the primary peer bonding involving the smallest groups, squads of eight to ten soldiers, or platoons of twenty to fifty most often associated with those 'bands of brothers'.[3] The specific chronological focus is the end of antiquity, especially the sixth century. Much scholarship on unit cohesion in other militaries has moved beyond a reliance on peer-bonding as the prime motivator in combat, but there has been little work done on unit cohesion of any sort, including horizontal peer bonding, in late antiquity.[4] This chapter remedies this by looking for evidence of bonding among groups of eight to ten men, as noted the size of group usually associated with primary group bonding. We then evaluate the evidence for groups of this size in action.

Scholarship

There has not been a great deal of research on cohesion in the late antique Roman military, and so our brief overview of the scholarship will concentrate on the Roman armies of the imperial era, from Augustus through the end of antiquity. Some scholars have focused on the morale of Roman soldiers. A good example of this is Lee's insightful chapter, which concentrated on morale in the Roman military, and which explored the evidence for primary groups, in his case *contubernales*, units of eight to ten men.[5]

Other scholars have examined cohesion more broadly. Smith's wide-ranging essay on military units throughout classical antiquity discussed aspects of the Roman era, including Rome's allegedly successful attempts at maintaining and increasing unit cohesion, with a particular emphasis on veteran colonies.[6] Goldsworthy's book, in which he adopted a face-of-battle approach in his analysis of Roman combat, devotes space to battle from the perspective of the unit, and the individual, with some discussion of cohesion.[7] Writing at about the same time, Southern offered one of the few discussions of cohesion in the context of late antiquity. She linked cohesion with *esprit de corps*, though her emphasis was particularly on morale, which Engen persuasively argues should be distinguished from cohesion.[8] Indeed, Southern refers specifically to a fourth-century episode described by Zosimus, which centres on the fearful flight of some Batavian soldiers.[9] Culham has noted how often unit cohesion has been examined since the World War II, and she draws parallels with the Roman context, though like some other contributions her emphasis is on the earlier period.[10] Some scholars have preferred to emphasize communities of soldiers. James, though he recognizes the significance of membership in small groups, and he even calls the members of these groups *contubernales*, stresses that we have no direct evidence for their bonding, a point to which we will return below.[11] Along the same lines, Driessen's focus is on Roman forts rather than the units, per se, though he makes a number of important points. He concentrates on groups larger than the primary ones – he specifically mentions secondary groups – that are usually the focus of studies of unit cohesion. Driessen ranges rather widely, however, for he discusses the relationship between comrades and superiors in group cohesion, stresses the importance of communal experiences, and even delves into *esprit*.[12] Some scholarship on the Roman imperial military has touched on primary peer bonding and unit cohesion, then, though as we have seen it is often tangential to treatments of other topics.

Background

Given that most of the chapters in the rest of this book deal with earlier eras, a little background on the Mediterranean world in the sixth century CE is in order. The setting for this chapter is the sixth-century East Roman Empire, and the chronological markers are 491 CE and 602 CE, the beginning of the reign of the emperor Anastasius, and the end of the reign of Maurice. The most useful sources for this chapter, which we discuss in the next section, are set within those years; hence, the markers. Like most periods of Roman history, the empire was beset with conflict, though more often than not the state was on the winning side of things. By this point the Roman Empire had split, with the west fragmenting into a number of smaller, though in some cases only a little less powerful, kingdoms, while the east maintained its territorial integrity to a significant degree. Most of the empire was eastern in respect to its geography and in relation to the Mediterranean.

It comprised a good part of the modern-day countries of Serbia, Bulgaria, Greece, Turkey, Syria, Lebanon, Israel, Palestine, Jordan, and Egypt. In the middle of that sixth century an attempt was made to recapture parts of the west, including North Africa, Spain, and Italy, with some success.[13] The military that is the subject of this chapter was not quite the institution that made Rome's armies famous, for though there was some continuity in terms of unit organization and even, possibly, in titulature, much had changed. Cavalry had become increasingly important,[14] armies were usually smaller,[15] and it seems that the express aim of most leading generals in the sixth century was the avoidance of pitched battle.[16]

Evidence Types

We have a great deal of evidence for the sixth-century East Roman Military. Our wide and varied evidence is composed of chronicles, chronographs, and histories, which provide military and political, ecclesiastical, or some combination thereof; narratives of events in the more recent and/or more distant past, written in a host of different languages; military manuals, in Greek and Latin, which tend to prescribe or describe military matters; papyri, primarily from Egypt though also from Jordan and Israel/Palestine, which include letters, lists, and assorted other documents that pertain to day-to-day minutiae; legal evidence, a huge portion of which was compiled under the orders of the emperor Justinian, and includes legal issues of more recent or more ancient import; inscriptions, which, though fewer in number than in earlier periods, include some important documents such as select edicts of the emperor Anastasius; and the physical remains of the sixth-century East Roman Military comprised of small items like beads, cooking utensils, and the very forts that many soldiers inhabited. Much of that material touches on armies and soldiers, in some capacity or other, and covers the gamut of military activities in the sixth century. Classicizing histories, like those of Procopius and Agathias, describe armies in action, while manuals like Maurice's provide both detail on organizational matters and comments on how armies should perform. The official documentation, the inscriptions, laws, and papyri, provide much of our evidence for the sixth-century military's organization, and so should use the appropriate nomenclature, to which we will return shortly. In sum, if there is evidence for cohesion formed by means of primary peer bonding, we should find it in this evidence.

Scholars of unit cohesion in more recent militaries have set the size of the primary group, the one involved in primary peer bonding, at between six and ten men, what is usually deemed in modern U.S. military parlance, for instance, as a group or squad. It so happens that for the earlier imperial era military there are Latin terms that would seem to correspond to the modern squad or group, and Lendon has looked at their usage in Latin inscriptions, with a particular emphasis on *contubernalis*, *commanipularis*, and *commilito*.[17] All three words can mean fellow soldier in some capacity or other, but there are subtle differences, with *commanipularis* usually referring to a member of the same infantry century (unit of 80 or so men within a legion); *contubernalis*, an emotional term that roughly means comrade but which might not refer to a soldier in the same unit; and *commilito*, perhaps the least affectionate of the three and which might not refer to a member of the same century.[18] *Contubernalis* is related to the term *contubernium*, which is a specific size within the Roman legion. That need not mean that we should exclude all or some of these terms from our discussion, however. While *commanipularis* and *commilito* would seem to be less useful, *contubernalis* and *contubernium* are extremely valuable, the former because of its relationship to the latter. Are these two terms, or something comparable, used in the sixth century?

We are reasonably well informed about the size and organization of the military in the sixth century,[19] if not quite as well informed as we are for the earlier imperial Roman armies that inhabited sites like Caerleon in Wales, where the fort's layout betrays the impact the *contubernium* would seem to have had.[20] In those earlier armies, the smallest unit was the *contubernium*, or a body of soldiers occupying a tent or room together (as suggested at Caerleon), which varied in size between eight and ten men, with ten being its size in later Roman history.[21] If we want to find small-group cohesion in action, it would make sense to look for evidence of these *contubernia* in the sixth-century evidence. Rather importantly, although the term *contubernium* was a feature of the earlier imperial era, the term, in some form or other, for instance as the Greek *dekarchy*, remained in use well into the sixth century. This is fortunate, for, as suggested, our knowledge of organizational matters in the sixth century is not quite as good as it is for the second century. It helps that, despite linguistic changes, both within the history of Latin and in terms of the increasing use of Greek in the Roman military, long a bastion of Latin use, military names and titles tended to persist for some time.

Lesser Evidence

Before we start looking at the potentially useful evidence, there are two important pieces of evidence for the age of Justinian that require consideration, Procopius' oeuvre and Justinian's *Corpus Iuris Civilis*. While they provide evidence for several aspects of waging war in the sixth century, as we will see they are not useful for tracking down peer bonds.[22]

Procopius operated in a tradition that favoured war and politics.[23] On the other hand, that tradition also emphasized an older form of Greek, and the general avoidance of non-ancient terms, which sometimes made for some wild circumlocutions.[24] Nevertheless, Procopius devotes considerable attention to war, and battles and sieges occupy a significant portion of the *Wars*.[25] One especially notable example is the Battle of Dara, fought in 530 CE between the forces of Rome and Sasanid Persia in what is today eastern Turkey.[26] This is one of the most detailed accounts of a battle we have from the sixth century, and throughout his description Procopius gives us all sorts of numbers and figures. For example, we are told that the Roman army numbered 25,000,[27] and Procopius even gives us totals for smaller contingents within that larger force, such as the 300 countrymen with the commander Pharas deployed on the right side of the main group,[28] and the 600 men to the right of Pharas under the commanders Sunicas and Aigan.[29] Unfortunately, besides some one-on-one single combat, these are the smallest divisions named in the account, and 300 is obviously far larger than the six to ten that scholars of small unit cohesion stress. Given, too, that the battle tends to be described in terms of grand strokes and movements, we would be hard pressed to find evidence of unit cohesion here. We just do not find descriptions of groups or squads in action. Even during Procopius' prolonged account of the 537/538 siege of Rome, in which the East Romans were locked within the city and besieged by Ostrogoths, and during which more than 60 sorties took place, evidence is lacking, since again, it is groups that number in the hundreds that are evident.[30]

Next, we turn to the legal evidence. Justinian was a master codifier of legal materials, or at least those who worked under him were. Important compilations include the *Codex*, *Digest*, *Institutes*, and *Novels*. We do find forms of *contubernium* and *contubernales* therein, however, the words could reflect other contexts. The living-together implied in those terms, is less evident in, say, Maurice's *dekarchy*, which refers to the size of the division, need not only apply to soldiers, but also to a husband and wife. In fact, in all those instances in which forms of *contubernium* and *contubernales* are used in Justinian's *Codex*, they are used pretty much exclusively in the wider context of marriage.[31] Indeed, a seventh-century history of Latin, written by Isidore of Seville, defines it as follows: "Living together' (*contubernium*) is an agreement to sleep together for a time; whence the term 'tent' (*tabernaculum*), which is pitched now here, now there'.[32] Like the evidence of the *Corpus Iuris Civilis*, the context here is marriage, with no military connotations implied.

In sum, it is hard to find examples of groups small enough for peer bonding in either Procopius or in Justinian's law codes. When Procopius does highlight individual elements, he usually focuses on generals, for Procopius' audience, largely members of the elite themselves, would be more interested in hearing about the great deeds of the leading figures of the day than the exploits of an average foot soldier, unless they were exemplary.[33] As for the legal material, while it provides plenty of evidence for military terminology, it is less useful for the military in action, and so this investigation.

Military Manuals

When we turn to military manuals, we find much more evidence of small groups. There are two texts worth discussing here: one, the *Epitoma Rei Militaris* of Vegetius, was written between about 375 CE and 450 CE, and was perhaps the most popular ancient text in the Middle Ages, though its current reputation could use improving.[34] The other, the *Strategikon* of Maurice, named after the emperor Maurice, who likely did not write it, dates to between 590 CE and 600 CE and is held in high regard by modern scholars.[35] Vegetius' work, usually considered prescriptive in nature, and so of lesser value, was written in Latin, while Maurice's, written in unpretentious and seemingly contemporary Greek, is usually considered descriptive in nature. In other words, where Vegetius' text prescribes what the Romans should do in combat, Maurice's describes what they did do in combat.[36]

Starting with Vegetius, there are fifteen places where he uses some form of the term. Where *contubernium* refers to the actual small section or group, *contubernales*, at least in theory, refers to the soldiers within that group. Vegetius usually uses that latter form. Unfortunately, it usually does not mean what we would like it to mean, and rather seems to be used much as it was in the earlier Roman funerary inscriptions discussed by Lendon.[37] There are several spots where *contubernales* means something like 'soldier',[38] 'ordinary soldier', or 'common soldier' (3.2.6, 3.10.4), or possibly 'private' (3.10.6). The remaining instances could refer to soldiers as comrades.[39] As regards *contubernium*, Vegetius uses a form of this term six times.[40] Indeed, there are points where Vegetius digresses to discuss select organizational nuances of the so-called ancient legion, something he refers to regularly, and which one scholar has identified with the legions such as they were during the Severan dynasty, some 200–250 years earlier.[41] In one such digression, Vegetius says the following:

> The centuries were themselves subdivided into *contubernia*. For every ten soldiers living under one tent, there was one in charge as 'dean' [*decanus*], now called 'section-captain' [*caput contubernii*]. The *contubernium* used to be called a 'maniple' because they fought in groups [*manūs*] joined together.[42]

Significantly, Vegetius has identified these sections and given their size. Again, though, it is worth stressing that the *contubernia* that Vegetius mentions here were historical sub-divisions within the Roman legion.[43]

A few sections later, Vegetius turns to communication on the battlefield, and, in particular, how to manage the confusion of battle.[44] Vegetius claims that, in the past, the soldiers used to paint different signs (*diversa ... signa*) on the shields so that soldiers (*milites*) would not stray from their comrades (*contubernales*). The catch, however, is that the signs painted on those shields were not for *contubernia*, but rather *cohortes* (cohorts), which were a much larger subdivision than tent-mate (*contubernium*). That passage, though dealing with soldiers in action and using seemingly relevant terminology, is of little use in uncovering unit cohesion at the primary group level. On the other hand, just a few sections later Vegetius makes this interesting statement:

> In this way, by means of the affection for the *contubernia*, the legionary cavalry honour their own cohorts even though horse are naturally inclined to be on bad terms with foot. By means, therefore, of this interweaving in the legions of all cohorts and of cavalry and infantry, one harmonious spirit was preserved.[45]

Not only did these small subdivisions, the *contubernia*, exist in Vegetius' eyes, but he identifies affection (*adfectio*) between these subdivisions. For Vegetius, then, while the language he uses varies considerably in meaning, he does identify units that fit the criteria of a primary group and he even goes so far as the claim that they had real affection for each other. On the other hand, Vegetius is concerned with the historical legion in this section, and so its applicability to the middle of the fifth century, when he was writing, let alone the sixth, is suspect.

As noted, Vegetius was writing a little before the period under investigation, even if we accept a later date for his work. More relevant for sixth-century matters is undoubtedly Maurice's *Strategikon*. The majority of the *Strategikon* is devoted to cavalry, with only the last book, of 12, covering infantry, which makes the text more useful to a specific arm within the military.[46]

The *Strategikon* was written in a form of Greek that likely had wide currency when it was composed, which was unlike Procopus' *Wars*, written in classicizing Greek. Maurice prefers a transliterated Greek form of *contubernium* (*kontoubernion*), the division itself, rather than the men that comprise one, *contubernales*. Maurice also uses the Greek equivalent, the *dekarchy*.[47] A *dekarchy* contained ten men.[48] It is worth noting that Maurice also discusses the so-called *pentarch*, a leader of five, in the same section, which implies an even smaller group, though he does not mention the group itself, only its leader.[49] Maurice implies that *dekarchies* act collectively, at least in some capacity. He claims (1.6) that the mission orders should be relayed through the commanders of these divisions, at least amongst the cavalry, and not before these groups have been put together. He also says that those in charge of these squads, *dekarchs*, should be the ones to administer punishments.[50]

On the other hand, Maurice's discussion implies that these divisions are put together for the express purpose of combat, which thereby limits the time needed for meaningful bonds to form. When he turns to who should fill these groups, he advocates for a mixture of young and old alike, a seemingly sensible assertion.[51] Again, however, this seems to be an ad hoc formation rather than something more permanent. Section 7.3.1 discusses organizing the *tagmas* into *kontoubernia*,[52] which suggests that these groups may not be a permanent thing, even if the larger focus of the section is strategy, which is a long-term matter. In the last book, Maurice calls the small groups of infantry *contubernia* (*kontoubernia*), which number 16 at most. The context, again, is combat, which also implies that they are an ad hoc creation. All in all, although Maurice notes the existence of groups of an appropriate size for unit cohesion to form within the sixth-century military amongst both the cavalry and infantry, the ephemeral nature of these groups did not leave much time for peer-bonding to occur.[53] The familiarity that is a hallmark of the primary group would have required much more time, the kind of time that the shared-living spaces of the earlier imperial military allowed for.[54]

Epigraphic Evidence

Epigraphic evidence plays an important part in our knowledge of the organization of the imperial era Roman military. Regrettably, the epigraphic habit, at least in Latin, falls away in the third century, though inscriptions do not dry up entirely in late antiquity. There are a few that discuss military matters, including the odd epitaph and honorary inscription.[55] Some of the most notable, and important, late antique era inscriptions for military history are the assorted decrees of Anastasius, inscribed in the first quarter of the sixth century. A few survive, and two significant examples are a decree from Cyrenaica pertaining to frontier soldiers and another from Perge, concerned with praesental soldiers. The most specific that the Cyrenaican inscription gets, in terms of military organizational terminology, is in its use of *arithmos* and *limitanei*.[56] The inscription from Perge, on the other hand, provides a wealth of detail about the organization and administration of the sixth-century military, with slab C, the third of the three slabs that comprise the inscription, devoted to the provisioning of the units discussed therein.[57] The unit found in the inscription, arguably a praesental legion, (a legion in the presence of the emperor that was based in, around, or fairly close to the capital),[58] was divided into even smaller groups. The terms used, presented in Latin, are not the forms we find in other contemporary or near-contemporary sources, and rather appear to be antiquarian.[59] What is more, some of the designations listed are familiar,[60] while others are not.[61] These designations are for specific kinds of soldiers and not the smaller subdivisions like a *contubernium*. There is no way of knowing from this document whether the legion was divided into groups as small as eight or ten men (*contubernia* and *dekarchies*) at this time.[62]

Papyrological Evidence

The penultimate category of evidence is the papyrological material. They cover various aspects of daily life, from personal letters between mothers and sons, or complaints to officers about suspect behaviour, to lists of supplies for military units. In the papyri from Nessana, a village in the Negev (Israel), a form of the word brother (*adelphos*) is used regularly, but it has to do with actual brothers, it seems, rather than any 'band of brothers'.[63] We do, however, find one mention of a *dekarch* in a papyrus from Nessana, namely P. Ness. 3.37, dated to 560–580; moreover, the papyrus seems to refer to allocations by *dekarchies*, the concomitant groups. Kraemer, the editor of the papyus, even says that the men were organized or grouped under their sergeants (*dekarchos*, *decurio*), a detail which point towards the existence of these small groups in Nessana.[64] This papyrus, a list of camels for the military unit at Nessana, seems to indicate that troops usually were organized into these squads at this particular location, even if we do not know what kind of unit, exactly, there was at Nessana.[65] On the basis of this papyrus, Kraemer surmises that the squadron was 200 men strong, with 20 sergeants. This would give it *dekarchies* of ten, which means that we have indirect evidence at Nessana for the equivalent of *contubernia*.[66] Given too that some of these men and camels were being dispatched to Egypt, we would then also seem to have indirect evidence of the squads, *dekarchies* or *kontoubernia*, in action.

If we shift from the Negev to Egypt, it so happens that forms of the word *contubernium* do appear in the papyri, and one in particular would seem, on the surface, to be particularly useful. The document in question, P. Oxy. 16.2046, dated to 546 CE, includes *contubernales* several times. It is a list of rations for the military and includes so-called Scythian soldiers (line 34) as well as a type of solder called a *bucellarius*, or biscuit-man (line 33). The *bucellarii*, found in comparative abundance in Egypt,[67] were private soldiers, perhaps even resembling mercenaries. Some served under big landowners in Egypt, others abroad under high-ranking generals like Belisarius. The term *contubernales* is found in a few places in the document along with the names of a host of leading individuals, presumably commanders. Although the lines that include the term *contubernales* do not specify exactly what kinds of soldiers they are, it is likely that these are *bucellarii*. This is because another papyrus, P. Oxy. 16.1903, which dates to the sixth century as well, is concerned specifically with *bucellarii* and names several individuals: Zemarchus (called 'the Bessian' in 1903), Boraides (also called 'the Bessian' in 1903), George, Paul, and Alexander, who are also found in P. Oxy. 16.2046. Given that the *bucellarii* based in Egypt tend to be associated with big landowners, and so were private soldiers who were unlikely to have been organized into any officially sanctioned types of groups,[68] there is no reason to believe that the term *contubernales*, as used in this papyrus, should be connected to any formally constituted *contubernia*. Rather, the *contubernales* would seem to be little more than comrades, with the term used in the same way that it often is in Vegetius' *Epitoma*. The Egyptian papyri, then, would also seem to have served up a dead end, while the Israeli papyri are considerably more promising.

Material Evidence

The last category to discuss is the material evidence, concentrating on the physical remains from, and within, the empire's fortresses. When soldiers were living together and regularly interacting with their peers, they would have had the opportunity to form the horizontal bonds that would seem to be the hallmark of unit cohesion. One fully excavated fortress is el-Lejjūn, located in Jordan, where the remains of the barracks have been uncovered. After an earthquake in 363 CE that damaged the fort, the original barracks were demolished and new ones were constructed. There were eight to ten rooms in the new barracks blocks, and the excavators associated these with the legion's *contubernia*. It is also worth pointing out that the consensus is that a legion was based at the fort throughout its history.[69] Those barracks were discovered to be full of domestic items, many of which would have been used in food preparation.[70] It seems likely that those who lived in the barracks did their cooking and eating together, not in the mess halls we usually associate with modern militaries. On the other hand, there is some evidence that women, and possibly children, were living with the soldiers in the barracks throughout the period under review.[71] In this scenario, the rooms were being shared by family members rather than by members of the same division, like a *contubernium*. If bonds were being formed, then, it is possible that they were with family members rather than fellow-soldiers, though the evidence is far from conclusive. There is, of course, no explicit reason based on the surviving physical evidence why soldiers could not have formed bonds with both.

Beyond Peer Bonding and Unit Cohesion

To this point the discussion has concentrated on looking for evidence of primary groups of eight to ten men, to indicate the peer bonding of the 'band of brothers' type that underlays unit cohesion. What we have seen is that some of that evidence is no use at all in trying to uncover groups that size. Procopius' descriptions do not include small groups operating in combat; the legal evidence, though it uses familiar terminology like *contubernium*, concentrates on marriage instead; and the epigraphic evidence is insufficient for this sort of discussion. Much of the papyrological evidence, at least that from Egypt, though it too contains seemingly familiar terms like *contubernalis* is in fact concerned with matters far less specific: comrades in general rather than comrades in specific eight- to ten-men groups. This is also true of much of Vegetius' language, with some exceptions, in whose *Epitoma* the word *contubernales* is usually employed to mean comrade, or something comparable; moreover, Vegetius' *contubernia* were historical divisions within his so-called *antiqua legio*. Maurice discusses groups of that size regularly, but as argued above that they were ad hoc groups, which would leave insufficient time for suitable peer-bonds to form based along organizational lines. Though Maurice certainly does provide evidence of primary groups in action, the ephemerality of their existence would seem to preclude the formation of those peer-bonds themselves. It so happens that there are hints that Maurice favoured other forms of cohesion, for while spending a significant amount of time discussing the procedures involved with meting out punishments, he stresses the importance of men staying in their place in line, and talks a bit about what should happen if men transgress these rules, so breaking up the unit's cohesion.[72] This is, of course, cohesion in the sense of a battle line staying together and not the small-group unit cohesion, formed by primary groups of eight to ten people, that is my focus.

All that is left is the allusion to small groups in the Nessana papyri. There seems to be little doubt that groups of ten men existed at Nessana, and the papyri suggest that they were often dispatched for specific tasks. In addition, based on what we know about the papyri and the wider community, there were plenty of opportunities for the men to form bonds in primary groups even if the soldiers lived with their families rather than their fellow soldiers.[73] The soldiers at Nessana, likely enrolled in a unit of *limitanei*, seem by and large to have been camel-riders and their tasks undoubtedly reflected their responsibilities: frontier matters, which in this region could include providing support for pilgrims.[74] While it is entirely likely that they saw combat, we cannot be certain.

Just as some have questioned the importance of primary-group unit cohesion in modern militaries there is good reason to doubt whether unit cohesion was a factor in the sixth-century East Roman military.[75] On the other hand, the apparent absence of evidence is not necessarily evidence of absence. We may simply lack the sources needed to discuss peer bonding. As we saw, there are instances where we find traces of evidence for the small groups usually associated with unit cohesion, the vaguest of allusions to these groups in action, and even more remarkably a hint of the peer bonding itself. As noted, Vegetius mentioned 'the affection for the *contubernia*', and how 'by means … of this interweaving in the legions of all cohorts and of cavalry and infantry, one harmonious spirit was preserved'.[76] But it comes within his book on the ancient legion and in his chapter on promotions within the military. The implication, then, is that this affection, what Milner translates as 'bonds' (i.e., bonding),[77] is created by the opportunities for promotion that existed with the ancient legion, and in fact seems to be for one branch of the legion, the cavalry, for another, the infantry. While Vegetius seems to demonstrate an understanding of bonding, it is not peer bonding, and it was not inherent to the military of his day, at least if we accept Vegetius' argument.

Although, at present, we lack evidence of horizontal peer bonding in the sixth century, that is not to say we have no evidence of other kinds of peer bonding. For one thing, sixth-century Roman soldiers were motivated, which would have implied unit cohesion on some level; moreover, although there were plenty of major battles that they lost, like Callinicum in 531 CE, there were plenty that they won, like Dara in 530 CE. To find combat motivation we need to look elsewhere.[78] Some ancient authors, like the anonymous author of the *Dialogue on Political Science*, Agathias, and Theophylact, stress that some soldiers were more bellicose than others.[79] In other cases, fear was an excellent motivator.[80] There were laws that punished those who fled from battle.[81] Soldiers who did so could be beaten with rods; however, the soldier who fled first suffered capital punishment.[82] Additionally, sixth-century sources are full of references to the impact of training on combat.[83]

Then there were other kinds of bonding, between soldiers and their organization or their unit, which could be expressed in terms of regimental pride. The reverence attached to the standards, which had their own shrines in forts like el-Lejjun, provide indirect evidence for this.[84] But we find standards serving this function in some narrative accounts too, like Procopius' of the Battle of Satala.[85] Maurice's detailed treatment of standards of various kinds also points towards their role as a means of fostering bonding between soldiers and their group.[86] Besides horizontal peer bonding there was also vertical peer bonding, especially that between a commander and his troops. Works like Procopius' *Wars* and Maurice's *Strategikon* both describe and proscribe how to be a general in sixth-century warfare and much of their work covers how to foster the relationship between the two.[87] With respect to the latter, the *Wars* is full of episodes in which the death of a commander brings, or nearly brings, an army to despair. Procopius argues vociferously that a general should not put themselves into a position that could bring them into danger, and so undermine their military objectives.[88]

Conclusion

At the end of this chapter and to get back to the point made by James noted earlier,[89] there is no clear evidence that primary-group unit cohesion was a factor in combat in the sixth century, which is in keeping with the scholarship for other militaries in other periods. We have evidence for bonding amongst larger groups, the cavalry and infantry of the ancient legion (Vegetius), and for smaller groups engaged in military actions (Nessana's *dekarchies*) at the end of antiquity, but we found few specific traces of primary-group bonding. More work is needed on sixth-century military communities in the sixth-century Roman military. Such work would allow us to understand better unit cohesion in all its axes and dimensions, and detailed evaluations of the roles of leadership, secondary group cohesion, shared experiences, 'swift trust', and training in the sixth century, following the approach advocated by Engen, would go a long way towards remedying this significant gap in the scholarship.[90]

Notes

1 I want to thank the editors and the anonymous reviewer for their valuable feedback. This paper was supported, in large part, by a Social Sciences and Humanities Research Council of Canada Insight Development Grant.

2 For example, Armstrong (2016), on the one hand, prefers 'military cohesion', while Birley (2002) and Parnell (2015) on the other, adopt some form of 'band of brothers' in their titles.

3 Siebold (2007): 289. See too Engen's (2016): 13–14, overview. For bonding I use the following definition: 'the social relationship, both affective and instrumental, of changeable strength (weak to strong) between service members and their group, organization, and service institution', Siebold (2007): 288.
4 For the Canadian army, see Engen (2016).
5 Lee (1996): 207–11.
6 Smith (1990): 161.
7 Goldsworthy (1996): 241–256.
8 Engen (2016): 14.
9 Southern (1996): 168; Zos. 4.9.2–4. Both Syvänne (2004): 261–71 and Janniard (2011) touch on some of the same issues as Southern, though they were writing much later.
10 Culham (2013): 258–9.
11 James (1999): 17.
12 Driessen (2005): 159.
13 The reconquest is described by Procopius in his *Wars* and Agathias in his *Histories*.
14 On the role of cavalry in the sixth century see Treadgold (1995): 93–8; Syvänne (2004); Petitjean (2014); though note the caution of Rance (2005).
15 On the size of late antique armies see Treadgold (1995): 43–64; Coello (1996); Nicasie (1998): 204–5.
16 See Luttwak (2009).
17 Lendon (2006).
18 Lendon (2006): 276.
19 See Parnell's (2017): 13–31, overview.
20 Note the comments of Speidel (2002): 133.
21 Note Tacitus' comments at *Ann.* 1.41 and *Hist.* 2.80.
22 See Jones (1964): chapter 17, on the legal evidence; Rance (2005) for Procopius and battle.
23 As for other historians that document the age of Justinian, Menander Protector's work is too fragmentary, Agathias adopts a similar practice to Procopius, Marcellinus and Malalas are far less interested in military matters, Evagrius regularly paraphrases Procopius, while Pseudo-Joshua was operating in Syriac, and so reflects a different outlook.
24 E.g., Cameron (1985): 115.
25 Whately (2016a), on how Procopius describes and explains battles and sieges.
26 For overviews of the battle see Greatrex (1998); Lillington-Martin (2007).
27 *Wars* 1.13.23.
28 *Wars* 1.13.19.
29 *Wars* 1.13.20.
30 Whately (2015b), on Procopius' use of numbers in battles and sieges.
31 E.g.: *CJ* 2.20.4; 3.21.2; 5.5.3 pr; 5.5.9; 5.27.5 pr; 6.27.3; 6.59.9; 7.16.3; 9.9.23 pr; 9.17.1; 11.8.3; 11.69.1 pr.
32 9.7.23, trans. Barney, Lewis, Beach, and Berghoff.
33 Whately (2016a): 219–24.
34 Charles (2007); Janniard (2008); Allmand (2011).
35 For an excellent survey of the late antique military manuals see Rance (2007): 343–8.
36 Maurice's Greek is filled with transliterated Latin terms and outright Latin commands, which makes identifying a form of the *contubernium* easier.
37 Lendon (2006).
38 Veg. *Mil.* 2.3.2; 2.10.3; 2.12.3; 2.14.5; 2.20.2; 2.14.5; 2.20.2; 2.20.6; 2.23.3; 3.2.6; 3.8.15; 3.10.4; 3.10.6.
39 Comrades: Veg. *Mil.*: 1.13.2; 2.13.3; 2.18.1; 3.4.5.
40 Veg. *Mil.* 2.7.5; 2.8.8; 2.13.6–7; 2.19.3; 2.21.4; 2.25.2.

41 Janniard (2011).
42 Veg. *Mil.* 2.13.6–7, trans. Milner.
43 Varro (*LL* 5.88, 6.85) makes the connection between the Latin for hand (*manus*) and maniple (*manipulus*), possibly why Vegetius has equated to two. If Vegetius is making his assumption based on etymology, he is joined by Isidore of Seville, referred to above. The maniple was much larger than a *contubernium*, possibly 120 strong, possibly more, though scholars disagree. Sabin, Van Wees, and Whitby (2007): 481, claim it was two centuries in size.
44 Veg. *Mil.* 2.18.1.
45 Veg. *Mil.* 2.21.4, trans. Milner, slightly modified.
46 This apparent emphasis on cavalry is not an indication that the use of infantry has all but disappeared by this time, but rather that the infantry was in such good shape that Maurice felt the need, at least at first, to concentrate on cavalry. See the comments and discussion of Rance (2005); (2007): 347–8.
47 Urbicius, who composed his brief *Epitedeuma* during the reign of Anastasius, uses the term *dekania* (4.26), as do Hesychius (E 6178) and the Suda (*Onom. Tact.* 5). Urbicius likely got his information from Aelian (*Tact.* 6.2.1) and/or Arrian (*Tact.* 6.2.1). On Urbicius, see Greatrex, Elton, and Burgess (2005).
48 Dennis (1984): 171, 173.
49 Maur. *Strat.* 1.6.
50 Maur. *Strat.* 1.8.1.
51 Maur. *Strat.* 2.6.
52 The *tagma* was equivalent to about 300 soldiers. See Dennis (1984): 173.
53 Treadgold (1995): 93, on the other hand, considers these dekarchies to be regular regiments.
54 On the time required for these bonds to form, see the comments of Siebold (2007): 289 and Engen (2016): 14–15. On the earlier Roman imperial context, Matthew (2015): 34.
55 Whately (2013): 111–12.
56 *SEG* 9, 356 and 414, Oliverio (1936): 135–63.
57 Onur (2014): 60, 62.
58 Onur (2012): 29; Onur (2016): 186–7.
59 Onur (2016): 185–6. It is not clear why these terms were used, and because the inscription has only recently been published it has not yet attracted due attention.
60 *Signiferi, optiones*; *vexillarii, imagniferi, librarii, mensores, tubiceni, cornices, bucinatores, armatores, beneficiarii, brachiarii, armatores*, and *munifices*. See Onur (2012): 60, 62. Slab B includes some other familiar designations, like *draconarius* (line 55, for instance). On many of these titles/ranks see Treadgold (1995): 87–93.
61 *Ordinarii, Augustales, Flaviales, Veredarii, Praeco, Torquati semissales, Clerici, Deputati.*
62 At the same time, the document leaves open the possibility that those soldiers of a similar rank might have formed primary or secondary groups of their own.
63 P. Ness. 3.15; 3.24.
64 Kraemer (1958): 114.
65 Whately (2016b).
66 Kraemer (1958): 114.
67 Sarris (2006): 162–75.
68 The Scythian soldiers mentioned in the same papyrus, by contrast, were likely government soldiers, and so to my mind are much more likely to have been organized into official groups.
69 Groot, Jones, and Parker (2006): 184.
70 McDaniel (2006): 295.
71 Whately (2015).

72 Maur. *Strat.* 1.8.2–1.8.3.
73 The evidence, such as we have it, is insufficient. In general, see the discussions of Rubin (1997); Urman (2004); Stroumsa (2008); Ruffini (2011); and Whately (2016b).
74 Whately (2016b): 131–3.
75 MacCoun, Kier, and Belkin (2006); Daddis (2010).
76 Veg. *Mil.* 2.21.4, trans. Milner, slightly modified. See above p. 000.
77 Milner (1993): 55.
78 Here I draw on Whately (2018), the complement to this chapter.
79 Anonymous, *Dialogue on Political Science* 4.19; Agathias 1.17.4; Theophyl. Sim. 3.7.9.
80 For a recent, provocative account of the role of decimation in motivating Roman soldiers of earlier eras, see Pearson (2019). She argues that fear did not play quite as important a role as is often argued in explanations of the role of decimation in Roman warfare.
81 *Dig.* 49.3.16.
82 *Dig.* 49.6.3.
83 Anonymous, *Dialogue on Political Science* 4.14–25; Agathias 2.1.2; Theophyl. Sim 3.12.7.
84 Lain and Parker (2006): 156.
85 *Wars* 1.15.15–16.
86 Maur. *Strat.* 1.2. See also Greet, this volume.
87 On generalship in the former work, see Whately (2016a).
88 Whately (2016a): 188–95.
89 James (1999): 17.
90 Engen (2016): 203–7. See, too, Siebold (2007); Armstrong (2016): 119. The necessarily brief discussions in the previous section have only scratched the surface on some of these issues, and more work would go a long way towards helping us understand what motivated soldiers in late antiquity, and the manifold bonds they formed.

Bibliography

Allmand, C. 2011: *The De Re Militari of Vegetius: The Reception, Transmission and Legacy of a Roman Text in the Middle Ages*, Cambridge.

Armstrong, J. 2016: 'The Ties that Bind: Military Cohesion in Archaic Rome'. In J. Armstrong (ed.), Circum Mare*: Themes in Ancient Warfare* (*Mnemosyne*, Supplement 388), Leiden: 101–119.

Barney, S. A., Lewis, W. J., Beach, J. A. and O. Berghof, trans. 2006: *The Etymologies of Isidore of Seville*, Cambridge.

Birley, A. 2002: *Garrison Life at Vindolanda: A Band of Brothers*, Stroud.

Cameron, A. 1985: *Procopius and the Sixth Century*, London.

Charles, M. B. 2007: *Vegetius in Context: Establishing the Date of the Epitoma Rei Militaris* (*Historia-Einzelschriften* 194), Stuttgart.

Coello, T. 1996: *Unit Sizes in the Late Roman Army* (BAR International Series 645), Oxford.

Culham, P. 2013: 'Imperial Rome at War'. In B. Campbell and L. A. Tritle (eds.), *Oxford Handbook of Warfare in the Classical World*, Oxford: 236–260.

Dennis, G. T. trans. 1984: *Maurice's Strategikon*, Phildelphia.

Driessen, M. 2005: 'Unifying Aspects of Roman Forts'. In J. Bruhn, B. Croxford, and D. Grigoropoulos (eds.), *Proceedings of the Fourth Annual Theoretical Roman Archaeology Conference*, Oxford: 157–162.

Daddis, G. A. 2010: 'Beyond the Brotherhood: Reassessing US Army Combat Relationships in the Second World War', *War & Society* 29: 97–117.
Engen, D. 2016: *Strangers in Arms*, Kingston, On/Montreal.
Goldsworthy, A. K. 1996: *The Roman Army at War 100 BC–AD 200*, Oxford.
Greatrex, G. 1998: *Rome and Persia at War, 502–532*, Leeds.
Greatrex, G., Elton, H. and R. Burgess, 2005: 'Urbicius' *Epitedeuma*: An Edition, Translation and Commentary', *BZ* 98: 35–74.
Groot, J. C., Jones, J. E. and S. T. Parker, 2006: 'The barracks in el-Lejjūn (Area K, L, R, and B.6)'. In S. T. Parker (ed.), *The Roman Frontier in Central Jordan: Final Report on the Limes Arabicus Project, 1980–1989*, Cambridge, MA: 161–185.
James, S. 1999: 'The Community of Soldiers: A Major Identity and Centre of Power in the Roman Empire'. In P. Baker (ed.), *Proceedings of the Eighth Annual Theoretical Roman Archaeology Conference*, Oxford: 14–25.
Janniard, S. 2008: 'Végèce et les transformations de l'art de la guerre aux IVe et Ve siècles', *Ant. Tard.* 16: 19–36.
Janniard, S. 2011: *Les Transformations de l'Armée Romano-Byzantine (IIIe – VIe Siècles apr. J.-C.): Le Paradigme de la Bataille Rangée*, unpublished PhD thesis, L'Atelier du Centre de Recherches Historiques.
Jones, A. H. M. 1964: *The Later Roman Empire*, Oxford.
Kraemer, C. J. 1958: *Excavations at Nessana, Vol. 3: The Non-Literary Papyri*, Princeton, NJ.
Lain, A. and S. T. Parker, 2006: 'The Principia of el-Lejjūn (Area A)'. In S. T. Parker (ed.), *The Roman Frontier in Central Jordan: Final report on the Limes Arabicus Project, 1980–1989*, Washington DC: 123–159.
Lee, A. D. 1996: 'Morale and the Roman Experience of Battle'. In A. B. Lloyd (ed.), *Battle in Antiquity*, London: 199–217.
Lendon, J. E. 2006: '*Contubernalis*, *Commanipularis* and *Commilito* in Roman Soldier's Epigraphy', *ZPE* 157: 270–276.
Lillington-Martin, C. 2007: 'Archaeological and Ancient Literary Evidence for a Battle Near Dara Gap, Turkey, AD 530: Topography, Texts and Trenches'. In A. S. Lewin and P. Pellegrini (eds.), *The Late Roman Army in the Near East from Diocletian to the Arab Conquest* (BAR International Series 1717), Oxford: 299–311.
Luttwak, E. 2009: *The Grand Strategy of the Byzantine Empire*, Cambridge, MA.
MacCoun, R., Kier, E. and A. Belkin, 2006: 'Does Social Cohesion Determine Motivation in Combat: An Old Question with an Old Answer', *Armed Forces & Society* 32: 646–654.
Matthew, R. 2015: Frater, Soror, Contubernalis: *Greedy Institutions and Identity Relationships in the auxiliary Military Communities of the Northern Frontier of Roman Britain in the First and Second Centuries A.D*, unpublished PhD thesis, University of Manchester.
McDaniel, J. 2006: 'The Small Finds'. In S. T. Parker (ed.), *The Roman Frontier in Central Jordan: Final Report on the* Limes Arabicus *Project, 1980–1989*, Cambridge, MA: 293–327.
Milner, N. P. trans. 1996: *Vegetius. Epitoma Rei Militaris*, Liverpool.
Nicasie, M. 1998: *Twilight of Empire*, Amsterdam.
Oliverio, G. 1936: *Documenti Antichi dell'Africa Italiana* II, Bergamo.
Onur, F. 2012: 'The Military Edict of Anastasius from Perge: A Prelininary Report'. In C. Wolff (ed.); *Le métier de soldat dans le monde romain*, Paris: 21–43.

Onur, F. 2014: *Monumentum Pergense. Anastasios'un Ordu Fermani*. Istanbul.
Onur, F. 2016: 'The Anastasian Military Decree from Perge in Pamphylia: Revised 2nd Edition', *Gephyra* 14: 133–212.
Parnell, D. 2015: 'Barbarians and Brothers-in Arms: Byzantines on Barbarian Soldiers in the Sixth Century', *BZ* 108: 809–926.
Parnell, D. 2017: *Justinian's Men: Careers and Relationships of Byzantine Army Officers, 518–610*, London.
Pearson, E. 2019: 'Decimation and Unit Cohesion: Why Were Roman Legionaries Willing to Perform Decimation?', *JMH* 83: 665–688.
Petitjean, M. 2014: 'Classicisme, barbarie et guerre Romaine: l'image du cavalier dans le monde Romain tardif', *AntTard* 22: 255–262.
Rance, P. 2005: 'Narses and the Battle of Taginae (Busta Gallorum): Procopius and Sixth-Century Warfare', *Historia* 54: 424–472.
Rance, P. 2007: 'Battle'. In P. Sabin, H. van Wees, and M. Whitby (eds.), *The Cambridge History of Greek and Roman Warfare, Volume II, Rome from the Late Republic to the Late Empire*, Cambridge: 342–378.
Rubin, R. 1997: 'Priests, Soldiers and Administrators: Society and Institutions in the Byzantine Negev', *Mediterranean Historical Review* 12: 56–74.
Ruffini, G. 2011: 'Village Life and Family Power in Late Antique Nessana', *TAPhA* 141: 201–225.
Sabin, P., van Wees, H. and M. Whitby (eds.) 2007: *The Cambridge History of Greek and Roman Warfare, Volume II, Rome from the Late Republic to the Late Empire*, Cambridge.
Sarris, P. 2006: *Economy and Society in the Age of Justinian*, Cambridge.
Siebold, G. L. 2007: 'The Essence of Military Group Cohesion', *Armed Forces & Society* 33: 286–295.
Smith, F. W. 1990: 'The Fighting Unit: An Essay in Structural Military History', *AC* 59: 149–165.
Southern, P. 1996: *The Late Roman Army*, London.
Speidel, M. 2002: 'The Framework of an Imperial Legion'. In R. Brewer (ed.), *The Second Augustan Legion and the Roman Military Machine*, Cardiff: 125–143.
Stroumsa, R. 2008: *People and Identities in Nessana*, PhD dissertation, Duke University.
Syvänne, I. 2004: *The Age of Hippotoxotai*, Tampere.
Treadgold, W. 1995: *Byzantium and its Army, 284–1081*, Stanford.
Urman, D. (ed.) 2004: *Nessana: Excavations and Studies I*, Beer-sheva.
Whately, C. 2013: 'War in Late Antiquity: Secondary Works, Literary Sources and Material Evidence'. In A. Sarantis and N. Christie (eds.), *War and Warfare in Late Antiquity: Current Perspectives*, Leiden: 101–151.
Whately, C. 2015: 'The Genre and Purpose of Military Manuals in Late Antiquity'. In G. Greatrex and H. Elton (eds.), *Shifting Genres in Late Antiquity*, Farnham: 249–261.
Whately, C. 2015b: 'Some Observations on Procopius' Use of Numbers in Descriptions of Combat in *Wars* Books 1–7', *Phoenix* 69: 394–411.
Whately, C. 2016a: *Battles and Generals: Combat, Culture, and Didacticism in Procopius' Wars*, Leiden.
Whately, C. 2016b: 'Camels, Soldiers, and Pilgrims in Sixth Century Nessana', *SCI* 35: 121–135.
Whately, C. 2018. 'Combat Motivation at the End of Antiquity'. In G. Greatrex and S. Janniard (eds.), *The World of Procopius*, Paris: 185–203.

9 '… They Were Routed': Cohesion and Disintegration in Ancient Battle

Louis Rawlings

The most significant test to cohesion that any military unit can experience is combat. It is an environment where the social and vertical bonds that tie soldiers together are subject to intense physical and psychological stresses. For some units, these pressures are overwhelming and lead to disintegration and rout, as soldiers turn and flee from the enemy. This chapter examines the reasons why some groups fail when they engage the enemy and seeks to understand rout through the lens of military cohesion. It will consider cohesion in three of its forms: horizontal peer cohesion, which is often regarded as deriving from social bonds of affiliation, vertical (or rank) cohesion, pertaining to the chain of command and the relationship between soldiers and their superiors, and task cohesion, reflecting the degree of commitment to the military operation.[1]

The ancients recognised the value of social bonds in combat. It was sometimes stated that men facing death in combat 'will gladly die in good company', or that men fought best where 'brother stands in rank beside brother, friend beside friend, lover beside lover'.[2] The Spartan poet, Tyrtaeus, explained that the most effective men in war were those who used words to encourage their neighbours.[3] Furthermore, it was recognised that when members of the unit acted under the scrutiny of peers and officers, social and vertical cohesion was strengthened. For Julius Caesar (*BCiv.* 1.67), the social eye encouraged people to stay in line, creating 'a sense of shame when all are looking on, and the presence of military tribunes and centurions also contributes much, and that it was by such considerations that troops are wont to be restrained and kept obedient'. Modern studies have confirmed that the more closely individuals identify with a group, the more they are able to endure the combat stress that they and their fellows are subject to.[4] In this sense, cohesion is a *combat enabler*, its role 'is to increase group solidarity or integration during times of stress, enabling individual members to withstand distressing circumstances, perform effectively, and contribute to group tasks and missions'.[5]

And yet many units have failed to withstand the violence of combat; indeed, some collapsed even before the enemy came into contact.[6] The stresses that bear hardest on cohesion, making a unit susceptible to disintegration

DOI: 10.4324/9781315171753-10

and rout, can be separated into those that are 'predispositional' (or 'contextual'), which are already present before battle, and those that are 'precipitating' (or 'trigger'), which arise during the engagement.[7] The former erode cohesion and make soldiers more susceptible to panic, while the latter provide the moments of existential threat that disrupt and overwhelm cohesion. Some pressures appear to act on specific aspects of cohesion, to dissolve the horizontal social ties within and between units, or to strike at the chain of command. Others damage task cohesion, affecting units' sense of purpose in ways that pose fundamental questions in the minds of soldiers, groups, and their local commanders: whether the assigned task is worth the risk, whether it can possibly succeed, or what to do when the plan and the battle appears to be going wrong.

This chapter will begin by outlining the predisposing factors that might affect a unit prior to the engagement and will reflect on how they work to make rout, flight and the loss of cohesion on the battlefield more likely.[8] This will be followed by a survey of a range of triggers that can cause units to lose their nerve and disintegrate under conditions of combat. Finally, there will be consideration of the extent to which cohesion persists during and immediately after a rout.

Predispositional Factors

If group cohesion is a combat enabler, then its absence may make a military force more vulnerable to the stresses of battle and prone to fragmentation. This is most clear in recently formed groups that consist of inexperienced or otherwise *ad hoc* assemblages of troops.[9] Horizontal cohesion often takes time to develop, so that makeshift forces, levies and militias who have been raised with the expectation of bringing only their 'three days' rations' may reach the battlefield as little more than bands of strangers.[10] Such forces had to rely on whatever social contacts existed in civilian life, although, for some members, past campaigning experience may have provided a repertoire of collaboration and cooperation that constituted a form of military cohesion.[11] The limits of such cohesion were very quickly reached; the Athenian phalanx, as Konijnendijk explains, was sometimes described by Greek writers as 'ill-prepared, skittish, insubordinate, weak and prone to fatal confusion … the moral pressure of the primary group and its values [w]as an unsatisfactory foundation for the Athenian hoplite's performance in battle'.[12]

Even in more experienced forces, cohesion could be fragile. Insubordination, dissent, and disaffection, could arise, which strained vertical ties. The unruliness of Fulvius' legions, who took little care in their deployment or in following their officers' commands, was claimed to be the root of their defeat to Hannibal at Herdonea (212 BC).[13] Such problems might be compounded when troops had specific reasons to lack confidence in their officers or in the commanders higher up the chain of command. At Crannon (322 BC), the Greek army 'melted away' because of 'their lack of obedience to their commanders, who were young and soft-hearted' (Plut. *Phoc.* 26.1). When a thousand mercenaries abandoned the campaign against the Carthaginians in 340 BC, they apparently 'declared that Timoleon was not of his right mind, ... marching against 70,000 enemy with only 5,000 foot and 1,000 horse' (Plut. *Tim.* 25.3; cf. Diod. 16.78.5-79.1). Disaffection was particularly acute during periods of civil war. When a state was riven by *stasis* and political discord, where loyalty was potentially mutable and individuals possessed conflicted emotions towards state or faction, then 'soldiers in the terror of civil strife are liable to consider their fears rather than their obligations'.[14] Suspicions of others' motives and issues of trust could undermine forces at critical moments, through fear that there may be traitors within the formation or in other parts of the army.[15] Indeed, Timoleon reportedly thought it beneficial that the thousand mercenaries had abandoned his campaign rather than showing their lack of commitment in battle (Plut. *Tim.* 25.4). General concern about the commitment, moral fibre and bravery of some individuals, was a common issue in many armies, though it was probably rare that forces could avail themselves of the stratagem of Iphicrates, who allowed those who were 'trembling and pale and showing every sign of fear', to go back to camp on a pretext of collecting their forgotten equipment, and while the 'cowards' were away, he called out to the rest 'now is the time for battle since we have got rid of our useless baggage!'.[16]

The deployment of an army consisting of units of inexperienced and veteran soldiers might also undermine broader cohesion, as the green troops may not only be more prone to panic, but the veterans may lack sufficient trust in their capabilities. It is perhaps little surprise that at Forum Gallorum, the veterans sent recently recruited cohorts back from the battleline, fearing their inexperience. Indeed, the veterans' instincts were confirmed when the raw cohorts did indeed break without engaging the enemy (App. *B Civ.* 3.67, 69). If lack of familiarity with the conditions of battle might make units unable to withstand the shock of combat, so too was a surfeit of the wrong sort of experience, acquired through past defeat. At Lampsacus (409 BC) the veteran forces of Alcibiades 'were unwilling to be marshalled with the troops of Thrasyllus; for they said that they had never been bested, while the others had just come from a defeat' (Xen. *Hell.* 1.2.15). The early books of Livy are replete with examples of Rome's Italian opponents: Sabines, Etruscans, and Samnites, whom he claims had become habituated to defeat at the hands of the Romans and were rarely capable of holding their ground.[17] Such statements display an obvious pro-Roman and ethnic bias; nevertheless, some units may have arrived on the battlefield already demoralised by their past experiences.[18] Furthermore, soldiers could be affected by the prior performance of other parts of the army, particularly in the days preceding a battle, when observing skirmishing between the camps. Here, one side might acquire a psychological disadvantage through repeated or heavy defeats by the enemy skirmishers or cavalry.[19] Such precedents ate away at the task cohesion of units, causing them to lose confidence not just in those recently defeated, but in the overall likelihood of success. Evidently, perceptions of the group's worth, which derive, in part, from its history and experience (or lack thereof), were significant aspects of cohesion.

In addition to concerns deriving from the experience, reputation and performance of units, soldiers' anxiety could be deepened by 'the empty alarms of war'. There were rumours that preyed on men's imaginations: 'trifling causes, either from groundless suspicions, sudden panics or religious scruples, have often produced considerable losses'.[20] Inexplicable panics, where the source was unknown and reaction irrational, were a potential danger to many armies, particularly at night or after military setbacks on a campaign. These required careful management, so that forces were not 'disturbed by sudden terror and be utterly lost', and Aeneas Tacticus recommended commanders

> station men in each watch of the night over every company or band, both on the flanks and in the centre, to take special care that, if they should perceive any disturbance coming on because of sleep or anything else, they may check it immediately.[21]

It is unclear how many armies followed such advice. Rumour was mightily potent in war and baseless or distorted reports of enemy movement and numbers (Thuc. 4.125.1; Polyb. 20.6.12) or the defeat of other friendly forces (Livy 9.44) could demoralise or cause panic in the ranks. Manifestation of the gods' will, through portents and omens might also undermine self-belief and evaluation of success in advance of battle.[22] While such phenomena could affect the whole army, sometimes the task cohesion of a unit could be particularly affected, as when a lunar eclipse caused the Gallic Aigosagai to refuse to continue in Attalus' campaign (Polyb. 5.78).

Bodily Discomforts

The impact of hunger, thirst, and fatigue on soldiers' capacity to act, think, understand, and remain aware of their environment can be considerable. Such stresses lead to a decline of mood, communication and cooperation, willingness to associate with the group, and cognition, as well an impairment of physical capacity.[23] Severe hunger, for instance, could weaken the resolve of individuals to remain in their groups or to continue with the army tasks, as they discover that their physical needs outweigh their desire to obey orders or even to defend themselves.[24] Long-term want of provisions could lower morale, posing questions in the minds of soldiers about the viability of the operation and the competence of the commanders in caring for their needs.[25] It contributed to pre-battle desertion and disaffection, but directly weakened the physical and psychological endurance of soldiers. At Vesontio, Caesar (*BG* 1.39–41) had to contend with a combination of his soldiers' fear of the Germans and their hunger; while at Aquileia, Herodian describes that the besieging 'soldiers were in a desperate position, short of everything ... these were the prevailing conditions of extreme privation and low morale'.[26] Even relatively short periods of hunger, where a force had been unable to breakfast before battle, could have notable impact, for instance, at Adrianople and Callinicum.[27] Operations in adverse weather conditions of snow, heavy rain, or heat could also take their physical and psychological toll on soldiers.[28] Indeed, Polybius notes that, at Ilipa, the heat of the day combined with empty stomachs made the Carthaginian troops feel faint, while at Trebbia the cold and damp of the winter river crossing and the absence of breakfast weighed heavily on the Romans.[29]

On the 6th of June 1944, during the D-Day landings, American troops found that the heaviness of their packs made it very difficult for them to assault Omaha beach. This was not just because the bulk slowed and encumbered the combatants, but because the burden seemed to drain their willpower: 'We were all surprised to find that suddenly we had all gone weak ... under fire we learned what we had never been told, that fear and fatigue are about the same in their effect upon an advance'.[30] A similar effect was present in 56 BC, during a charge made by the Venelli, who carried heavy bundles of brushwood a mile uphill aiming to fill the ditches of Sabinus' camp. When they arrived, overburdened and fatigued, they were immediately routed by the Roman attack (Caes. *BG* 3.19). Units who entered battle after long marches, night disturbances, or having suffered continual harassment over several days or weeks prior to battle were similarly undermined by fatigue.[31]

The psychological and physical pressures leading up to battle, then, could often have an important role in undermining cohesion during combat. Fatigue, hunger, weather, fear of the enemy, or lack of confidence in fellows, degrade an individual's capacity to function and to cooperate with others in a militarily effective way, making the units vulnerable to disintegration. Of course, many of the predisposing factors discussed here could affect not just individual units, but whole armies. Indeed, when several units suffer from the same conditions, other things being equal, it might not be unreasonable to suggest that all such units would succumb at the same point. However, the realities could always be somewhat messier, with variations in context potentially creating difference in staying power. For instance, on a long march, units in the van might benefit from better opportunities to forage, obtain firewood and clean water, and might act selfishly towards latecomers: 'the men who had arrived early and were keeping a fire would not allow the latecomers to get near it unless they gave them a share of their wheat or anything else they had that was edible' (Xen. *An.* 4.5.5). Such inequalities potentially undermined cohesion, making some individuals and groups physically or psychologically more vulnerable to the 'nonsocial, nonadaptive and nonrational behaviour of panic'. However, the predispositional factors outlined here did not, in themselves, normally force a unit to collapse in battle. Panic and rout required the presence of other, more immediate, triggers.[32]

Precipitating Triggers in Combat

Soldiers entering combat have always had to confront fear. They are suspended between the 'physical fear of going forward and the moral fear of turning back'.[33] On the day of battle, tension mounted, as troops marched out, deployed, and then began to manoeuvre. The perception of threat and of fear of injury and death increased as they came closer to the enemy. Then came the shock of combat itself and the trauma of fighting, where emotion, adrenaline and physical exertion reached its apogee.[34] It is in such circumstances that cohesion played its essential role as a combat enabler, where the ties of the collective exerted a 'moral fear' on individuals not to let their companions down or to be exposed to the shame and punishment that might follow exhibitions of cowardice. At such moments, spatial proximity and the physical sense of companions advancing played a key role in reassuring individuals of their continued affiliation and attachment to the group.[35] And yet, combat produces pressures that can transform individual fear into the mass panic of rout. These are stress triggers capable of disrupting interpersonal relationships, making soldiers more susceptible to losing the sense of group belonging, and undermining the feeling of obligation, service and sacrifice to the collective: 'then the members are less able to resist the stresses of battle and are more predisposed to behave as individuals (i.e. non-socially) and become more concerned with self-survival than with the survival of the group as a whole'.[36]

Ardant Du Picq, with much insight drawn from personal experience as a serving officer of the French army, observed that:

> ... man has a horror of death ... the mass always cowers at sight of the phantom, death. Discipline is for the purpose of dominating that horror by a still greater horror, that of punishment or disgrace. But there always comes an instant when natural horror gets an upper hand over discipline, and the fighter flees. "Stop, stop, hold out a few minutes, an instant more, and you are victor! You are not even wounded yet, - if you turn your back you are dead!" He does not hear; he cannot hear any more. He is full of fear.[37]

The panic of soldiers during battle was widely recognised by the ancients.[38] The Greek orator Gorgias described it in the following terms:

> ... whenever belligerents in war put on their bronze and iron equipment, some for offence, others for defence against enemies, if sight beholds this, it is alarmed and it alarms the soul, so that often panic-stricken men flee future danger [as if it were already] upon them. For however strong the habit of obedience to law, it is banished because of the fear prompted by such sight, which makes one heedless both of what is customarily honourable, and of the good that comes from victory. Some

> who have seen dreadful things have lost their minds in the present time; thus, fear extinguishes and drives out understanding.[39]

Men are overwhelmed by their sensations: the sight of equipment and of 'dreadful things', that flood the imagination with thoughts of danger.[40] Modern commentators have also noted that the sensory overload that comes from sudden traumatic shock, an overwhelming sense of threat to survival, real or imagined, which was capable of tipping fear into desperation.[41] In such circumstances, as some members of the group recoil or flee, fear spreads and then, Xenophon observes, 'it is very difficult to find men willing to stay in place, when they see some of their own side in flight'.[42] At times, this led to a dramatic collapse of the line at the very first contact, or even before it; the so-called 'Tearless Battle' (368 BC) being just one of a number of such occasions.[43] Rout could spread very quickly from unit to unit along a battleline, so that at Delium and Leuctra, when one section gave way, other contingents were swiftly discouraged and also routed.[44] Similarly, during the battle of Second Philippi, when the first line cohorts were driven back, Brutus' second and third lines fled.[45]

Casualties

One of the most direct assaults on cohesion comes from the impact of combat casualties. Witnessing the effect of violence upon others' bodies close-up can bring home to the individual the tangible possibility of death. Furthermore, the demise or incapacitation of officers disrupts the vertical chain of command, while the loss of close companions directly strikes at the ties of social cohesion. We might expect casualties to be a primary determinant in the failure of unit cohesion. However, the interaction of casualties and cohesion appears to be complicated and non-linear. Cohesion appears to act in such a way that when individuals are close to attachment figures (i.e. those with whom they share a social or vertical bond), then even quite extreme levels of threat do not precipitate flight.[46] Instead, the usual outcome is intense affiliate behaviour. In the face of danger, cohesion intensifies as the men seek reassurance in the proximity of their peers and the leadership of their officers, which makes a unit more resistant. Commanders who seem particularly warlike, effective and reassuring in stressful contexts facilitate this process – for instance, Clearchus, who was 'ready by day or night to lead his troops against the enemy, and self-possessed amid terrors …' so that

> … in the midst of dangers, the troops were ready to obey him implicitly and would choose no other to command them; for they said that at such times his gloominess appeared to be brightness, and his severity seemed to be resolution against the enemy, so that it appeared to indicate safety.
> (Xen. *An.* 2.7, 11)

The death of such individuals, however, could immediately make the environment appear to be less safe and the consequent disruption of vertical cohesion could cause a unit to cease to function. The loss of the oversight and direction of command could provide an excuse for soldiers to act in their own self-interest.[47] It could quickly trigger the disintegration of units and spread rout through battlelines.[48] Even if soldiers remained cohesive, they might nevertheless take counsel among themselves to decide on escape.[49] Damage to vertical cohesion also occurred when ancillary members of a command group, such as standard-bearers or trumpeters, were killed. The loss of a standard could cause confusion and terror in the ranks (Procopius *Wars* 1.15.15–16). This was a symbolic blow to the identity of the unit, a disruption to communication and dissemination of orders, and a removal of a focal point for movement, resistance, and rally.[50]

Awareness of mortality is a powerful generator of stress and fear. In combat it is derived from a combination of the casualties sustained by peers and a perceived threat to an individual's own life.[51] The shock of initial casualties falling upon a unit has the potential to create a psychological crisis, in which individuals struggle to master the acute fear of imminent death.[52] Loss of comrades intensifies the sense of danger and simultaneously weakens the social obligations to remain. As the social networks begin to be struck down throughout the unit, these ties that bound the unit together are weakened or severed.[53] However, although the loss of comrades potentially wore away at social cohesion, paradoxically, it might spur men to fight harder to protect the fallen from being despoiled, or the wounded from falling into enemy hands.[54] Indeed, Socrates stated that 'in many instances, where men have tried to rescue a comrade or relative in battle, they too have been either wounded or killed'.[55] Social cohesion can sustain individuals, who might find it 'good to die' with 'buddies', such as the *hetairoi* of Cimon who all fell at Tanagra.[56] Indeed, studies of modern armies have shown that highly cohesive units are able to endure quite extreme levels of casualties before they cease to function.[57] In ancient battles, when highly motivated contingents encountered one another, there could be prolonged combat, with notably heavy casualties, which might even result in the destruction of the units where they stand, for instance, at Forum Gallorum, where Octavian's praetorians fought to the death.[58] Such resistance might also manifest when escape seemed impossible and 'the only hope of safety for the defeated was to expect no safety' (Veg. *Mil.* 3.21). At Telamon (225 BC), where the Gallic army was sandwiched between two Roman forces, some 40,000 fought to the death, while at the Great Plains (204 BC) the Celtiberian mercenaries 'had no hope of safety in flight, owing to their ignorance of the country, nor could they expect to be spared if captured'.[59]

Some enemy commanders recognised the danger of forcing defeated men into such a hopeless situation and considered that a better approach was to leave avenues of escape.[60] Indeed, in most ancient battles, defeated units readily availed themselves of such opportunities. The result was that many units cracked after sustaining relatively light (or indeed minimal) casualties in melee. Whereas, at Telamon, a reported 57% of the entire army of Gauls were killed; a study of ancient hoplite battles concluded that the defeated tended to suffer, on average 14%.[61] It is difficult to separate from these figures the losses that came before and after a unit routed. The casualties inflicted on victors averaged around 5%, which may suggest the impact on units while they were still fighting (since fewer of the victors are likely to have routed or pursued). Sabin therefore speculates that either one side was simply more effective at killing in melee or, more likely, both sides suffered limited overall losses during the actual melee, but, for the losers, many more might have been killed during pursuit, when men were panicked and no longer resisting.[62] If this latter was the normal situation, then the cause of rout, for many forces, appears due not so much to the casualties they sustained, but to other physical and psychological pressures.

Combat Exertions

We have already seen that preexisting fatigue can affect cognitive and co-operative capabilities. Combat itself was an extremely physically and psychologically draining experience that could further degrade combatants' sensemaking and ability to collaborate.[63] It is difficult to estimate the degree to which the duration of the combat experience impacted the endurance of soldiers, since it is often unclear how long combat lasted. Pritchett's survey of hoplite battles suggests that some clashes were over in a matter of minutes, with rout coming very soon after the first contact, while others lasted 'a long time'.[64] In several recorded cases, it appears that prolonged bombardment from arrows, sling-stones and other missiles sapped energy and morale, the 'spindles' that a Spartan survivor of Sphacteria sardonically quipped about, did not discriminate between the cowards and the brave (Thuc. 4.40). The Galatians at Ancyra (189 BC) and the Roman forces harassed by Massinissa in Spain (211 BC) were worn down by such skirmishing until they were gripped by 'alarm and confusion ... anxiety and fear'.[65] Just the physical burden of holding weapons and shields in a fighting or protective stance, could be a considerable drain. In melee, where men fought hand to hand, they could quickly become exhausted by the physical acts of stabbing, slashing, and parrying.[66] It is possible that fatigue may have played a role even when battles appear to have been decided in the immediate contact, or at least in a few brief minutes of savage and brutal fighting, before one side gave way. However, some longer battles appear to be characterised by an ebb and flow along the battle line. In such cases combat appears punctuated by moments of pause and short withdrawals by combatants in order to recover some energy, and in which there were alternating flurries of combat and lulls between opposing units.[67] In such engagements, units might be pushed back step by step over time (Thuc. 4.96; Polyb. 11.24.7) or might attempt to disengage from the immediate fighting, to reform and rally some distance behind their original position (Thuc. 4.43; Xen. *Hell.* 3.5.20).[68] In such contexts, as the fight wore on, exhausted soldiers finally became incapable of summoning the energy to attack, or indeed to continue to fight.[69] Then soldiers were unable to resist the next push forward of their enemy, and decided just to keep running rather than reform, their cohesion disintegrated through the repeated stress of these battlefield movements.[70]

Vegetius observed that 'when a tired man enters battle with one who has rested, or a sweating man with an alert, or one who has been running with one who has been standing, he fights on unequal terms'.[71] Enemy reinforcements therefore provided fresh danger: owing to their vigour, alertness (perhaps a capacity to see opportunities to strike or to initiate parries), and psychological insulation from the previous trauma of the fighting.[72] Even victorious units could be routed if they encountered a fresh force willing to stand up to them; they quickly discovered they had no reserves of energy. Hirtius' fresh army near Mutina overpowered Anthony's troops, worn down by their earlier exertions that day in defeating Pansa's army at Forum Gallorum.[73] Similarly, the Athenians on Epipolae, during their pursuit, suddenly broke when they encountered a steady Boeotian unit.[74]

The Opposition

Just as the capabilities of one's own side (or perceived lack thereof) could undermine confidence, the expectations of the prowess and qualities of the enemy played an important role in the psychological pressures acting on the group. It struck at task cohesion, making soldiers less sure that they would be capable of victory. The intimidation felt among Caesar's troops on hearing about the prowess and stature of Ariovistus' Germans required the commander's intervention, while the Macedonians who saw the effect of deathblows delivered by Roman weaponry 'realized in a mass panic what weapons and what men they had to fight'.[75] An acute appreciation of the enemies' qualities may have been at the root of many sudden collapses of phalanxes confronted by Spartans. It is likely that the units who were deployed against them could not tolerate the knowledge of their reputation, already expecting that 'to fight the Spartans would be a fearful matter'.[76] Furthermore, aural or visual intimidation on the day of battle might escalate tension, deterring and unsettling a force. An enemy war cry could create panic and terror in the battleline.[77] The physical appearance, for instance, of Spartans or Gauls, or the nature of an enemy advance – be it wild and ferocious or disciplined and precise, could contribute to an overload of senses and emotion, and cause terror to snap the reins of cohesion.[78]

The performance of an enemy may sometimes exceed expectations, so that an engagement which might have been regarded as relatively easy or straightforward becomes much more difficult owing to their bravery or skill.[79] If noticeably more effective than anticipated, their actions can serve to undermine task cohesion– challenging perceptions of how the battle may be going.[80] Overconfident troops, who attack incautiously, are particularly vulnerable to the shocks delivered in such a form (Thuc. 2.11.4). Thucydides related how 'the Argives rushed forward on their own wing with the careless disdain of men advancing against Ionians who would never stand their charge, but were defeated by the Milesians with a loss little short of three hundred men' (Thuc. 8.25). When expectations (the sense-making or groupthink) of the coming encounter come into conflict with the facts of the situation, confidence can turn to despair.[81]

Besides the psychological expectations of the capabilities of the enemy, their tactical capabilities posed practical and direct threats to the cohesion of units. Their deployment may be quicker, leaving the opposition still attempting to organise themselves when the attack commences.[82] It would have created acute unease among units who suddenly realised that they had insufficient time to prepare for the engagement. At Amphipolis, when Brasidas saw the Athenians were manoeuvring in disorder, he realised that 'Those fellows will never stand before us, one can see that by the jostling of their spears and heads. Troops which do as they do seldom stand a charge'. Indeed, when his force suddenly charged, the Athenians routed, 'terrified by their own disorganisation and astounded by his audacity' (Thuc. 5.10.5–6).

Enemy tactics can have a great impact on the cohesion of units, especially when these produce mismatches in fighting methods or tactical approaches.[83] Superior mobility and missile power can torment even the most cohesive and committed force. In battles characterised by tactics of evasion and skirmishing of light troops against heavy infantry, the latter's formation was vulnerable to fragmentation as it experienced the repeated push-me-pull-you of ineffectual pursuit and rally-back, usually under a hail of missiles and insults, which fatigued the unit, and created the conditions for psychological collapse.[84] Such enemy tactics exploited the frustration and exasperation of slower-moving and heavier armed soldiers, whose expectations and capabilities were tailored towards fighting hand to hand. Without their own covering force of skirmishers, such as the slingers that were hastily organised for the Ten Thousand,[85] the inability to get to grips with the enemy demonstrated impotence, precipitating a crisis in task cohesion, as the *raison d'etre* of the fighting unit (to inflict melee damage in combat) was undercut. At Carrhae, a mixed force of Roman cavalry and infantry became cut off on a hill by Parthian horsemen, and 'all alike were hit with arrows, bewailing their inglorious and ineffectual death' (Plut. *Crass.* 25.10), while at Olympus, the Galatians became

> blinded by rage and fear, they did not know what to do … when wounded by missiles, flung from afar by unseen enemies, and having no-one against whom to make a blind assault, they ran mindlessly into each other, like speared wild beasts.
>
> (Livy 38.21.7–8)

The Spartans at Sphacteria and Lechaeum suffered the evasive skirmishing of the enemy light troops and were unable to act as they preferred. Instead, the Spartans were rendered almost powerless by the missile bombardment and seemingly incapable of influencing events.[86] At Sphacteria, there had been no chance for heroes to be distinguished from cowards, as the enemy missiles, 'spindles', fell upon them (Thuc. 4.40.1).

Tales of the Unexpected: Sensemaking and Disintegration

The unexpected often can have a devastating impact on cohesion by directly assaulting the sensemaking of a unit. Sensemaking can be broadly understood as the way soldiers and officers interpret the course of events and the conditions that they find themselves in, based on information they possess, their expectations and experience, their perception of the situation, and their capacity to interpret and adapt quickly to changed circumstances. As the battle unfolds, soldiers' perceptions may not necessarily accord with what is actually occurring, and their misunderstanding or lack of awareness can persist until it is too late. When the reality of the situation finally becomes clear, it is often an unpleasant surprise. Suddenly the universe is turned upside down as the members of the group try to make sense of the changed circumstances. Without swift guidance or reassurance this can lead to confusion and panic. Whether it be through ambush or sudden attack, unexpected enemy reinforcements, flanking, or even the *force majeure* of environmental phenomena (such as sudden rain or lightning), the 'worst fortune' can lead to a catastrophic crisis in confidence that units may not recover from.[87]

We can see this situation unfold during the night attack on Epipolae in 413 BC. The Athenians' initial successes were very swift, and they become overconfident in their advance. However, after a unit of Boeotians had broken part of their forces, other elements took a long time to understand the changed situation. In the darkness and unfamiliar terrain, failures in sensemaking manifested in several ways: firstly, situational disorientation as to the positions of the enemy and the directions they were facing, secondly, confusion and disorder as advancing friendly units blundered into those who were retreating and even fell to fighting one another, while others revealed their watchword to the enemy, and thirdly, the terror of those who were intimidated by the sounds both of the enemy war-cries and those of their allies. The shock of the changed situation and the inability to make sense of the realities of their position caused units to panic and led to a disastrous flight down from the heights.[88]

Some of the most significant impacts on sensemaking occur when a unit finds itself unexpectedly flanked (either after a break-through in the line, or envelopment of the formation) or is surprised by previously unknown forces.[89] The physical contact from an unexpected direction (particularly if the unit is already engaged by another enemy force), leads to additional combat stress from exertion, fatigue and casualties, combined with a fear among the combatants of being trapped and crushed together.[90] These stresses combine to incline members of the group to attempt to escape such a dangerous situation through flight.[91] The unexpected appearance of enemies, however, may sometimes be more shocking than the reality, and fear takes a grip that is beyond the magnitude of the situation – thus at Delium, the manoeuvre of Boeotian cavalry to emerge from behind a hill terrified the Athenians into thinking a second army was approaching (Thuc. 4.96.5).

When there was a disconnection between the desires and orders of the commanders and the understanding of the unit of its situation and chances of success, then vertical cohesion could be at risk. Members of the unit may perceive a tactical or strategic mistake by the commanders, whether from their own ignorance of the broader course of events or awareness of the battle plan, or from a more acute understanding of the difficulties and dangers of their own situation, which they might perceive as hopeless or militarily nonsensical.[92] This, for the unit, compromises its ability to complete the task. It can lead to the unit becoming isolated from command, drawing it into taking decisions for itself.[93] Units rely on realistic orders from commanders to be effective, even if these are as basic as 'go there' or 'charge them'; their absence or surfeit can place the unit in a crisis of sensemaking. Polybius (18 25.6) observed that the Macedonian left wing at Cynoscephalae, having nobody to give them orders, were unable to fight effectively against the Romans, while at Syracuse, Thucydides (6.72.4) noted the opposite, that too many orders from too many officers were being given during the battle. Both situations were regarded as critical in the defeats suffered by those forces.

The Persistence of Cohesion: Affiliative Behaviour in Rout

The various predispositional and precipitating factors identified in this discussion strike, in various ways, at the bonds of cohesion and have the potential to turn the collective into a myriad of disassociated, panicked, individuals. The scattering of soldiers in pell-mell flight is often facilitated by the topography of battlefields, particularly those fought on open terrain, such as plains, which offer many unobstructed routes of escape. Such topographical conditions would appear to enhance the possibility of unit disintegration. Other battlefields, however, offered more challenging conditions for escape, through the presence of defiles, rivers, walls or other difficult to transit terrain. These sometimes resulted in crowding, where the routers were forced together and where units might be forced to remain together, despite their panic, and although placed in great peril of destruction by the enemy or the crush of their own numbers.[94]

However, even when the terrain might be favourable for dispersal and every-one-for-themselves behaviour, it is striking that some groups retained a semblance of formation or, at least, its members remained in close proximity, during their escape. In such cases, it appears that individual and group preservation elided, and unit cohesion persisted, though in opposition to the wider military objectives of the army and its integrity. Battle, then, could create a tension between social cohesion (the companions, buddies, fellow unit members) and vertical cohesion (the unit in its relation to the chain of command).[95] There could also be dissonance between social cohesion and task cohesion, where the unit no longer operated to further the task, but gave way to collective self-preservation. Indeed, in some cases, failure was caused because the primary group itself permitted the cessation of fighting after, perhaps, a token resistance or a more carefully scripted kind of posturing, preliminary to surrender or retreat, and which was often an outcome of pre-battle demoralisation or disaffection.[96] In what appear to be rational acts of collective action, those Peloponnesian allies at Leuctra who were 'not unhappy' with the failure of the Spartans on the right, themselves quickly gave ground (Xen *Hell.* 6.4.15, 16). However, even during outbreaks of mass panic, particularly in response to sudden, unforeseen, battlefield events, a degree of affiliative behaviour was often displayed.[97] Modern studies have observed that it is quite normal at such times for individuals to move towards friends, family and known affiliates, including superiors. In theatre-fires and mass shootings, more often than not, families and friends have been observed to spend valuable time seeking one another out, even at the expense of immediate escape and personal safety; such behaviour has also been frequently identified in military contexts.[98] Evidence for such conduct in antiquity is admittedly thin, but tent-mates such as Socrates and Alcibiades, even though serving in different arms at the battle of Delium, might cohere in the chaos of flight.[99] More frequently reported are cases of soldiers remaining within their immediate group or unit during rout, so that they flee collectively in a particular direction rather than dispersing willy-nilly. Livy, for example, reports that at the battle of Luceria (320 BC), those Samnites who escaped, headed for Apulia 'in scattered parties', until they rallied together and returned to the town of Luceria (Livy 9.13). At Forum Gallorum, the Martian legion, having been driven off the battlefield in rout, reached Hirtius' camp together and, despite its fatigue, possessed sufficient self-awareness to reform outside (and away from other panicked units), and prepared to resist again (App. *B Civ.* 3.69). Collective action, to present a fighting or a defensive withdrawal, was normally regarded as an effective way to enhance the chance of survival during the slaughter that often accompanied a rout.[100] For the most part, the winners found it easier to kill the unresisting, scattered, and panicked than to target such groups.[101]

Nevertheless, the main explanation for the affiliative behaviour of routers, particularly those who are paying scant attention to resisting the enemy any further, but merely trying to escape, appears to be the comfort provided by the familiar. Clinging to recognizable preexisting relationships (friends, family, comrades, group leaders) is a reassurance in crisis, when expectations have been shattered and individuals find themselves in an unfamiliar and terrifying context, for which their experience has little or no precedent to draw upon.[102] In addition, during escape, soldiers normally move towards familiar and psychologically comforting locations that are perceived as safer environments – camps, muster points, terrain that appears to offer refuge (such as hills or forests), or previously travelled landscapes that present recognisable avenues of escape.[103] Of course, such locations often are places where others are likely to gather, so that dispersed members of units might reassemble, to rally or collectively decide to resist, to surrender, or to move away. Indeed, Vegetius (*Mil.* 3.25) recommends that a general whose army has routed should try to retrieve the situation by visiting such locations in order to collect sufficient troops, either for a fight-back or a safer retreat. On the whole, the escape to 'havens of comfort' aligns with affiliation responses: the familiar contexts of the camp, the unit, family, and/or home.[104] Perceptions, of course, were not always correct. There were numerous instances where 'safe' environments provided little more than temporary respite; for when the victors targeted or surrounded such locations, their safety was revealed to be phantasmagorial.[105]

Conclusion

The aim of this study has been to open discussion and identify factors that acted adversely on unit cohesion, before and during battle. By necessity, it has only been able to scratch the evidential surface and to engage briefly with some of the salient scholarship. Nevertheless, it is hoped that it will stimulate further analysis and debate about the most significant threats to cohesion posed by ancient conditions of battle.

Some of the clearest indicators of a unit's vulnerability to battlefield rout relate to the ways in which its cohesion is put under pressure. Men in battle are normally forced to rely on one another for their survival, and the weaker the bonds of trust and expectation soldiers have in one another, the less they are able to achieve this. In addition, physical hardships, hunger, and fatigue, which emphasise to individuals their own personal discomforts, undermine their capacity to cooperate, to focus on tasks, or to obey orders, and additionally affect their decision making and perception of their environment. Such men are susceptible to the magnifications of fear caused by sudden violence, the realisation that an enemy really is capable of killing them (even if it may be statistically not so likely, at least for those behind the foremost ranks), and particularly by unexpected and unsettling events beyond their control. These can drive men to assert what little agency they have left by trying to escape their position in the line. When enough individuals react in this manner, then the unit disintegrates. Yet, while rout signals the failure of task and vertical cohesion, even as soldiers flee, they may remain close to their comrades, seeking, through such affiliation, safety in the havens of comfort that these relationships represent. Thus, even as cohesion fails, it can persist; many who flee retain their sense of the collective, and it appears that at least some disintegrated units are able to re-form, once the terrors of the battlefield are left behind.

Notes

1 For fuller definition and discussion of these aspects see Hall's Introduction to this volume, p. 1-5.
2 Good company: Veg. *Mil.* 3.21, a view expressed also by the stoic Seneca *Contr.* 9.6.2. Brothers, friends, and lovers: Onos 24.
3 Tyrt. Frg. 12.18–19. Cf. Plut. *Cato Min.* 73.3. On the importance of talking to companions in combat: Marshall 1947: 137; Hanson 1989: 100; King 2006: 505.
4 Griffith and Vaitkus 1999: 40–1; cf. Marshall 1947; Wong *et al.* 2003.
5 Griffith 2007, 141.
6 For example, in 368 BC at the 'Tearless Battle' where the Arcadians fled from the charge of the Spartans; further examples in Pritchett 1974: 203–5, see also below n.43.
7 Schultz 1971.
8 Schultz 1971: 4–8.
9 Inexperienced troops lacking cohesion are sometimes described by Livy as *tumultuarius*, *incondita turba*, or *multitudinis*: Livy 22.44, 25.1, 3, 26.40.16, 29.28, 30.8; they tended to fare badly when confronted by more seasoned and bonded forces. In a similar vein, Alcibiades described Sicilian soldiers as ὅμιλος – a mob – and that their internal discord would be an extreme liability for them in combat (Thuc. 6.17.4).
10 Three-day's rations: Aristoph. *Ach.* 197–8; *Vesp* 242–4, *Peace* 312. While some scratch-built forces did evidence high cohesion and commitment, it is often because they were designed for specific military tasks: various volunteer or chosen bands who brought individual personal commitment and commitment to the mission, which offset deficiencies in social cohesion. The best *ad hoc* forces combined elements of task cohesion with some pre-existing social bonds. The one hundred infantry and one hundred cavalrymen selected by Mago to form an ambush-party on the eve of Trebbia (Polyb. 3.71.5–9; Livy 21.54), may have been selected for the mission because of military qualities known to the commander, but strikingly each man then picked ten others, suggesting the importation of small-group affiliations (Hall and Rawlings, this volume, p. 88).
11 Hanson 1989, Crowley 2012: 23–5, 43–7, Konijnendijk, this volume p. 12.
12 Konijnendijk, this volume, p. 22.
13 Livy 25.21, other examples cf. 22.21; 26.2, 28.24; also note Thuc. 3.72, 6.72.
14 Caes. *BCiv.* 1.67; cf. the alleged poor performance of disaffected plebeians at Eretum and Algidus (Livy 3.42, 8.36).
15 Megarian confidence was undermined by fear of pro-Athenian collaborators (Thuc. 4.68.2); according to Theopomp. *Hist.* FGrH 115 F 93, Cleon, after an unknown engagement, charged the Athenian cavalry with desertion (*leipostration*); Christ 2006: 117, n. 60; Bugh 1988: 111–4.
16 Polyaen. *Strat.* 3.9.1; cf. Theophrastus *Characters* 25.4 where the coward, as combat begins, returns to camp on that very pretext. At Leuctra, the Thespian commitment was doubtful and so Epaminondas allowed them to leave the army (Paus. 9.13.8); Pritchett 1974: 205.
17 Livy 2.6, 8.16, 8.38, 10.41; 23.40, *Cf.* Caes. *BG* 6.24.
18 Polyb. 14.8.7–8: 'At the first encounter the Numidians gave way before the Italian horse and the Carthaginians before Massinissa, their courage having been broken by previous defeats'.

19 Goldsworthy 2000: 56, 371 n. 36; Rawlings 2016: 232.
20 Caes. *BCiv.* 3.72; Schultz 1971: 6–7; Wheeler 1988.
21 Aen. Tac. 26.1 (a frequent occurrence), 6 (men disturbed and undone), 12 (setting a panic over-watch). On other measures taken by commanders see Wheeler 1988: 175–6.
22 E.g. Macedonians before Pydna were demoralised by a rumoured interpretation of a lunar eclipse (Polyb. 29.16.3); a Gallic army was terrified by thunder at Delphi (Paus 10.23.1–2); Pritchett 1974: 91 ff; Wheeler 1988: 166; Hanson 1989: 105.
23 Murphy 2002: 27.
24 Obedience: Xen. *An.* 4.5.7–9. Self-defence: Thuc. 7.84.4–5.
25 Xen. *Hell.* 6.2.19: At Corcyra (373 BC), some officers (*lochagoi*) pointed out to Mnasippus that 'it was not easy to keep men obedient unless they were given provisions' and when they marched into battle, 'they were all dispirited and hostile to him'. Lee 2019.
26 Herodian 8.5.6, 8. Lee 2019: 282–3.
27 Adrianople: Amm. Marc. 31.12.13. Callinicum: Procopius *Wars* 1.18.37.
28 Xen. *An.* 4.5.3–20; Plut. *Pomp.* 35.2; Plut. *Crass.* 25.9; Plut. *Ant.* 46.4–47.4, Amm. Marc. 31.12.10–13.19; van Lommel 2013: 166.
29 Ilipa: Polyb. 11.24.5; cf. Livy 28.15. Trebbia: Polyb. 3.72.3–5; cf. Livy 21.54.
30 Staff Sergeant Thomas B. Turner, Company M, 116th U.S. Infantry Regiment, Barnthouse 1986: 28.
31 Marches: Caes. *BCiv.* 2.41; Amm. Marc. 31.12.10; Veg. *Mil.* 3.11; Onos. 6.9. Night disturbances: Livy 25.34. Harassment: Livy 25.34. See Murphy 2002: 27.
32 Schultz 1971: 8.
33 Thompson 1884: 662.
34 The psychological aspects of the deployment, advance and combat are well described by Hanson 1989.
35 Shils and Janowitz 1948: 288.
36 Schultz 1971: 6, cf. Schultz 1964. His view that less cohesive units are more vulnerable to group terror reflects the discourse in military psychology that both Du Picq (1921) and Freud (1922) had observed: that combat stress can create a situation where individual self-interest overcomes regard for and loyalty to the group. The chain of causation has been a matter of debate: whether panic causes a unit to collapse (McDougall, 1920) or whether the disintegration of the unit leads to panic (Freud, 1922, 43–5 ff.). It has been the latter position that has had the greatest impact on the discourse of military cohesion.
37 Du Picq 1921: 2.6.
38 Hanson 1999, 96–106.
39 Gorg. *Hel.* 11.16–17.
40 Ustinova and Cardeña 2014; other symptoms of fear: see Hanson 1989: 98 ff; Crowley 2012: 162 n. 70 collects relevant examples.
41 Heidenreich and Roth 2019; cf. Anders 2011, 233–5 on fear in Roman battle.
42 Xen. *Hell.* 7.5.24; Hanson 1989: 160; Christ 2006: 104.
43 Xen. *Hell.* 7.1.31; cf. Thuc. 5.72.4; Caes. *BCiv.* 2.34; Pritchett 1974: 203–5; Hanson 1989: 102–3; Sabin 2000: 13; Rawlings 2007: 94; Konijnendijk 2018: 179.
44 Delium: Thuc. 4.96.6. Leuctra: Xen. *Hell.* 6.14.4.
45 App *B Civ* 4.128; other egs: App. *B Civ* 3.69; Caes. *BCiv.* 2.34; 3.64, 69.
46 Mawson 2005: 100.
47 Schultz 1971: 11, cf. Freud 1922: 48.
48 Deaths of centurions: Livy 25.19 (M. Centennius Paenula); Caes. BCiv. 1.46; 3.64. Of a *toxiarchos*: Thuc. 3.98.1. A brief and random selection of other slain officers: Thuc. 3.98.4 (Procles); Xen. *Hell.* 4.3.8 (Polycharmus), 4.8.29 (Therimachius); 5.1.12 (Gorgopas); Livy 29.2 (Indibilis); Dio 49.20.3 (Pacorus); see Peithis 2021: 45.

49 E.g. Caesar *BCiv.* 3.69. By contrast, where chains of command possessed some resilience, and there remained sufficient availability of subordinate or natural leaders within the unit, then vertical cohesion could be maintained. The Spartan chain of command typically provided such resilience, so that at Sphacteria, the unit continued its resistance despite the losses of senior commanders, it was a surviving junior officer who finally took the decision to surrender (Thuc. 4.38.1).
50 See Anders, this volume.
51 Kolditz 2006: 656–7.
52 Caes. *BCiv.* 3.51; Kagan 2006, 128.
53 Konecny 2014: 34–5.
54 Alcibiades' related that, at Potidaea, 'none other than he [Socrates] saved me. He refused to abandon me when I was wounded and saved both me and my armour' (Pl. *Symp.* 220e); cf. Plut. *Alc.* 7.3. Compare the fight over Pacorus' body (Dio 49.20.3).
55 Pl. *Alc.* 115b; Anderson 2005: 279.
56 Plut. *Cim.* 17, though Cimon himself was prohibited from taking his place in the line, due to his exiled status, nevertheless his panoply was taken into the fight.
57 Wainstein 1986: 26.
58 App. *B Civ.* 3.69.1; cf. the Carthaginian Sacred Band at Crimissus (Diod. Sic. 16.80.4, on which see Hall and Rawlings, this volume p. 83), or several Thespian phalanxes (e.g. Thuc. 4.96; Xen. *Hell.* 4.2.20; note the illuminating discussion in Hanson, 1999).
59 Telamon: Polyb. 2.30.9. An additional 10,000 were captured. The original size of the army had been 50,000 infantry and 20,000 cavalry (Polyb. 2.23.4), thus around 20,000, possibly mostly cavalry, may have escaped. Great Plains: Polyb. 14.8.9.
60 Frontin. *Str.* 2.6.5; 4.7.16; Veg. *Mil.* 3.21.; Procop. *Wars* 4.12.22–4.
61 Telamon, see above n.59. Ancient casualty figures are notoriously untrustworthy, but the conditions at Telamon, where an army was surrounded, could well have produced such a proportion of casualties. Hoplite battles, see Krentz 1985; cf. Sabin 2000: 5–6, 11 who notes a similar rate of 5% for victors in Roman battles.
62 Sabin 2000: 6, 11.
63 Caes. BCiv. 2.41; App. *B Civ.* 3.68.
64 Pritchett 1985: 47–51. Hanson 1989: 25, 153, 219, 227, who estimates an hour for most hoplite battles. Duration of Roman battles: Sabin 2000, 4–5.
65 Ancyra: Livy 38.27. Spain: Livy 25.34, cf. Frontin. *Str.* 2.5.25, Polyaen. *Strat.* 6.38.6, Zon. 9.3.
66 Livy 25.12; Goldsworthy 1996, 224–7; Hanson 1989: 56, 68, 77. If the initial attack of Gauls could be endured then, our sources claim, they became easily fatigued: Polyb. 2.35.6; Livy 22.2; 27.48; Dio Cass 12.50.3.
67 E.g. Hdt. 9.61–3; Plut. *Pel.* 32; App. *B Civ.* 3.68, 69, 4.128.
68 The literature modelling combat is extensive and precludes full treatment here. This discussion draws on Sabin's model (1996; 2000) of Roman warfare, but also Rawlings 2007: 93–7 for hoplite battle; note Wheeler 2011, for a more sceptical analysis, and Konijnendijk 2018, for a review of the various models.
69 Konijnendijk (2018: 188) also sees rout as a gradual process. He emphasises the role of group psychology in contributing to defeat.
70 Sabin 2000: 14–16.
71 Veg. *Mil.* 3.11; cf. Onos. 6.9.
72 Reinforcements' insulation from fear: du Picq 1921: 1.2. The impact of reinforcements: Thuc. 2.79.4–5, 4.43.4, 5.9.8; Xen. *Hell.* 1.3.6, 4.3.6–7, 4.5.17; Polyb. 10.6.11; 18.21–2; Caes. *BCiv.* 3.51, 69, 93, 94.
73 App. *B Civ.* 3.70; cf. Polyaen. *Strat.* 3.9.54 (Iphicrates).

74 Thuc. 7.43.7. Other examples: Thuc. 4.25.11; Livy 38.25; App *B Civ* 3.70; Veg. *Mil.* 3.25.
75 Germans: Caes. *BG* 1.39. Roman weapons: Livy 31.34; cf. warlike reputation of Illyrians: Thuc. 4.125.1.
76 Lysias 16.17; cf. Thuc. 4.34.7; Hanson 1989: 98–9; Konijnendijk 2018: 179–80.
77 Thuc. 4.34.2; 7.44.6; Polyb 11.16.2, 18.25.1; App *B Civ* 3.68; Plut. *Crass.* 26.3; Veg 3.18.
78 Physical appearance of Spartans: Xen. *Lac Pol.* 10.3.8; of Gauls: Diod. Sic. 5.30.5, 5.31.1; Livy 21.28, 38.21. Advance of Gauls: Polyb. 2.33.2; Strabo 4.198; Aristot. *Eth. Nic.* 3.5 b 28; of Spartans: Plut. *Lyc.* 22. 2–3; of Macedonians: Arrian *Anab.* 1.6.1–3.
79 E.g. Carthaginian mercenaries at Adys 256 BC (Polyb. 1.30.11); Tarentine Mercenaries (Polyb. 11.14.1); Afranius' cohorts (Caes. BCiv. 1.44).
80 Thuc. 6.70.1; Caes. BCiv. 3.93 (unexpected ferocity of fourth line at Pharsalus).
81 Xen. *Hell.* 4.5.12, 17, at Lechaeum, where Spartan initial contempt for Iphicrates' peltasts turned to utter despair: Konecny 2014: 33–4.
82 Thuc. 5.9.6–7, 5.10.6; Xen. *Hell.* 7.5.22; Polyb. 10.39.6; Caes. BCiv. 3.63; Veg. *Mil.* 3.18.
83 Quality: Polyb. 11.24.7; Livy 30.8. Tactics: Livy 26.4; Caes. *BCiv.* 3.93 at Pharsalus combined cavalry and a fourth line of infantry against cavalry.
84 E.g. Sphacteria (Thuc. 4.34.2), Spartolus (Thuc. 2.79.6), Aetolia (Thuc. 3.94.3), Syracuse (Thuc. 7.79.5), Lechaeum (Xen. *Hell.* 4.5.15–16), Mount Olympus (189 BC, Livy 38.21); Ilerda (Caes *BCiv.* 1.44), Bagradas (against Juba, Caes. *BCiv.* 2.41), Carrhae (Plut. *Crass* 24–5, 27).
85 See Marshall, this volume p. 35-7.
86 Sphacteria: Thuc. 4.34–6. Lechaeum: Xen. *Hell.* 4.5.14–17; see Konecny 2014: 33–6.
87 See also Anders, this volume p. 114. Ambush : Livy 7.6. Sudden attack: Livy 25.18. Enemy reinforcements: Livy 28.13. *Force majeur*: Caes. BCiv. 7.32, when a sudden phenomenon occurs, such as a flash of lightning; cf. e.g. Frontin. *Str.* 1.12.12; note Thuc. 6.70, where the inexperienced were affected by rain, thunder, and lightning, while the veterans fought on, but, two years later, the already demoralised Athenians regarded sudden thunder and rain as an ill-omen (Thuc. 7.79.3); this after the disastrous delay caused by an eclipse (Thuc. 7.50 ff). The 'worst fortune': Lys. 16.15; Christ 2006: 108–9, sees this as an excuse for rout.
88 Thuc 7.44; Foster 2018: 114–117.
89 Flanking: Livy 44.41. Ambush: Thuc. 3.108; Xen. *Hell.* 5.1.12; Polyb. 3.71.5–9 (cf. Livy 21.54); Polyb 3.82–3; 3.104–5. Livy 25.21. On surprise attacks see Krentz 2000.
90 Xen. *Hell.* 7.5.24; Caes. *BCiv.* 2.34, 3.69.
91 Thuc. 5.10.8–9; Xen. *Hell.* 4.2.21–2; Polyb. 10.31.3, 18.26.3–5; Caes. BCiv. 3.69, 94; Procop. *Wars* 4.12.19–21.
92 Soldiers calling on commanders to desist from certain course of action or offering advice: e.g. Thuc. 5.65.2; Plut. *Phoc.* 25.
93 Thuc. 5.72 provides the example of two officers, Aristocles and Hipponoidas, who refused to obey Agis' command to shift their units' position in the battleline during the advance.
94 Walls: Xen, *Hell.* 4.4.11–12 (Corinth). Gates: Livy 10.5 (Rusella), Caes. BCiv. 2.35 (Utica). Ditches: Polyb. 11.16.2, 11.17.6 (Mantineia). Defiles: Thuc. 3.94.1–98 (Aetolia); Procop. *Wars* 4.12.19–25 (Mt. Bourgaon). Cliffs: Thuc 7.44.8 (Epipolae). Field boundaries: Thuc 1.106.1 (Megara). Water: Polyb 3.72 (Trebbia).
95 During rout, vertical cohesion can survive if it reconfigures to prioritise the survival of the unit. Sometimes junior officers might attempt to extract their units from defeat (Caes. BCiv. 3.95: centurions and tribunes). Indeed,

commanders might be forced to go with such a flow, ordering a general retreat in an attempt to retrieve control or, at least, direct the course of the rout or panic (e.g. Thuc. 1.63; Caes. *BCiv.* 2.35, 2.41). However, commanders might then run the risk that they would be blamed and, in some communities, prosecuted, for ordering such a retreat, see Christ 2006: 104.

96 Wessely 2006: 278; e.g. Livy 25.15: 'The Thurians, an ill-disciplined crowd, disloyal to the side on which they fought, were at once put to flight'.

97 Mawson 2005.

98 Mawson 2005: 101.

99 Plut. *Alc.* 4.4, 7.2; Pl. *Symp.* 219e, 220e, 221a-c; Anderson 2005: 286.

100 Pl. *Symp.* 221a–b; Christ 2006: 105.

101 A point made by Alcibiades (Pl. *Symp.* 221c); cf. Tyrt. Fr. 11.10–14; Caes. *BCiv.* 3.95; App. *B Civ* 4.128; see Hanson 1989: 177–84. Antony's pursuit of coherent knots of routers after Philippi (App. *B Civ.* 4.129) indicates a particular desire to shatter all current and future resistance.

102 Weick 1993, 633–4; van Epps 2008.

103 Mawson 2007: 174; Milne 2012: 29. A few egs: Thuc. 4.44.1, 4.96.7, 5.10.9–10; Polyb. 11.24.7; App. *B Civ.* 3.70, 4.128; Procop. *Wars* 1.15.16; Veg. *Mil.* 3.25.

104 Mawson 2005: 102, 2007: 174.

105 E.g. Thuc. 1.106.1–2; Polyb. 3.84.13–14; Caes. *BCiv.* 3.95. App. *B Civ* 4.129.

Bibliography

Anders, A. 2011: *Roman Light Infantry and The Art of Combat: The Nature and Experience of Skirmishing and Non-Pitched Battle in Roman Warfare 264 BC–AD 235*, PhD Dissertation, Cardiff University.

Anderson, M. 2005: 'Socrates as hoplite', *Ancient Philosophy* 25: 273–289.

Barnthouse, Cpt. C. 1986: 'Infantry in action: Sustainability', *Infantry Magazine* 76 (1): 27–32.

Bugh, G. R. 1988: *The Horsemen of Athens*, Princeton.

Christ, M. R. 2006: *The Bad Citizen in Classical Athens*, Cambridge.

Crowley, J. 2012: *The Psychology of the Athenian Hoplite: The Culture of Combat in Classical Athens*, Cambridge.

Du Picq, A. 1921: *Battle Studies: Ancient and Modern Battle*, trans. Col. J. N. Greely and Maj. R. C. Cotton, Gutenberg ebook 7294, published online 2005.

Foster, E. 2018: 'Military defeat in fifth century Athens: Thucydides and his audience'. In J. H. Clark and B. Turner (eds.), *Brill's Companion to Military Defeat in Ancient Mediterranean Society*, Leiden: 97–122.

Freud, S. 1922: *Group Psychology and the Analysis of the Ego*, trans. J. Strachey, London and Vienna.

Goldsworthy, A. K. 1997: 'The othismos, myths and heresies: The nature of hoplite battle', *War in History* 4: 1–25

Goldsworthy, A. K. 2000: *The Punic Wars*, London.

Griffith, J. 2007: 'Further considerations concerning the cohesion-performance relation in military settings', *Armed Forces & Society* 34: 138–147.

Griffith, J. & M. Vaitkus, 1999: 'Relating cohesion to stress, strain, disintegration, and performance: An organizing framework', *Military Psychology* 11 (1): 27–55.

Hanson, V. D. 1989: *The Western Way of War*, London.

Hanson, V. D. 1999: 'Hoplite obliteration: The case of the town of Thespiae'. In J. Carman and A. Harding (eds.), *Ancient Warfare: Archaeological Perspectives*, Stroud: 203–217.

Heidenreich, S. M. and J. P. Roth, 2019: 'The neurophysiology of panic on the ancient battlefield'. In L. L. Brice (ed.), *New Approaches to Greek and Roman Warfare*, Hoboken: 127–138.

Kagan, K. 2006: *The Eye of Command*, Ann Arbor.

King, A. 2006: 'The word of command: Communication and cohesion in the military', *Armed Forces & Society* 32: 493–512.

Kolditz, T. A. 2006: 'Research in *in extremis* settings: Expanding the critique of "Why They Fight"', *Armed Forces & Society* 32: 655–658.

Konecny, A. 2014: 'Κατέκουψεν τὴν μόραν Ἰφικράτης. The battle of Lechaeum, early summer 390 BC'. In N. V. Sekunda and B. Burliga (eds.), *Iphicrates, Peltasts and Lechaeum* (*Monograph Series 'Akanthina'* 9). Gdansk: 7–48.

Konijnendijk, R. 2018: *Classical Greek Tactics: A Cultural History*, Leiden.

Krentz, P. 1985: 'Casualties in hoplite battles', *GRBS* 26: 13–21.

Krentz, P. 2000: 'Deception in archaic and classical Greek warfare'. In H. van Wees (ed.), *War and Violence in Ancient Greece*, London: 167–200.

Lee, A. D. 2019: 'Food supply and military mutiny in the Late Roman Empire', *Journal of Late Antiquity* 12: 277–297.

MacDougall, W. 1920: *The Group Mind*, New York.

Marshall, S. L. A. 1947: *Men Against Fire*, New York.

Mawson, A. R. 2005: 'Understanding mass panic and other collective responses to threat and disaster', *Psychiatry* 68: 95–113.

Mawson, A. R. 2007: *Mass Panic and Social Attachment: The Dynamics of Human Behavior*, Burlington.

Milne, K. 2012: 'Family paradigms in the Roman Republican military', *Intertexts* 16: 25–41.

Murphy, Lt. Col. P. J. 2002: *Fatigue Management During Operations: A Commander's Guide*, Puckapunyal.

Peithis, S. C. 2021: *Tactical and Strategic Communications in Ancient Greece, Fifth Century BC*. PhD thesis, University College London.

Pritchett, W. K. 1974–1985: *The Greek State of War, Part II* (1974), *Part III* (1979), *Part V* (1985), Berkeley, Los Angeles, and London.

Rawlings, L. 2007: *The Ancient Greeks at War*, Manchester.

Rawlings, L. 2016: 'The significance of insignificant engagements: Irregular warfare during the Punic Wars'. In J. Armstrong (ed.), Circum Mare*: Themes in Ancient Warfare*, Leiden: 204–234.

Sabin, P. A. G. 1996: 'The mechanics of battle in the Second Punic War'. In T. J. Cornell, N. B. Rankov and P. A. G. Sabin (eds.), *The Second Punic War: A Reappraisal*, *BICS Supplement* 67: 59–79.

Sabin, P. A. G. 2000: 'The face of Roman battle' *JRS* 90: 1–17.

Schultz, D. P. 1964: *Panic Behavior*, New York.

Schultz, D. P. 1971: *Panic in the Military*, Charlotte.

Shils, E. A. and M. Janowitz, 1948: 'Cohesion and disintegration in the Wehrmacht in World War II', *Public Opinion Quarterly* 12: 280–315.

Thompson, D. L. 1884: 'With Burnside at Antietam'. In R. U. Johnson and C. C. Beul (eds), *Battles and Leaders of the Civil War*, Vol II, New York: 660–662.

Ustinova, Y. and E. Cardeña, 2014: 'Combat stress disorders and their treatment in Ancient Greece', *Psychological Trauma: Theory, Research, Practice, and Policy* 6: 739–748.

Van Epps, Maj. G. 2008: 'Relooking unit cohesion: A sensemaking approach', *Military Review, November–December 2008*: 102–110.

van Lommel, K. 2013: 'The recognition of Roman soldiers' mental impairment', *Acta Classica* 56: 155–184.

Wainstein, L. 1986: 'The relationship of battle damage to unit combat performance', *Institute for Defence Analyses Paper P-1903*, Alexandria, VA.

Weick, K. E. 1993: 'The collapse of sensemaking in organizations: The Mann Gulch Disaster', *Administrative Science Quarterly* 38: 628–652.

Wessely, S. 2006: 'Twentieth-century theories on combat motivation and breakdown', *Journal of Contemporary History* 41: 269–286.

Wheeler, E. L. 1988: 'Ῥολλὰ κενὰ τοῦ πολέμου: The history of a Greek proverb', *GRBS* 29: 153–184.

Wheeler, E. L. 2011: 'Greece: Mad hatters and march hares'. In L. L. Brice and J. T. Roberts (eds.), *Recent Directions in the Military History of the Ancient World*, Claremont, CA: 53–104.

Wong, L., Kolditz, T. A., Millen, R. A. and T. M. Potter, 2003: *Why They Fight: Combat Motivation in the Iraq War*, Carlisle, PA.

Index

Page numbers followed by "n" indicate notes

For Product Safety Concerns and Information please contact our EU representative GPSR@taylorandfrancis.com
Taylor & Francis Verlag GmbH, Kaufingerstraße 24, 80331 München, Germany

www.ingramcontent.com/pod-product-compliance
Ingram Content Group UK Ltd.
Pitfield, Milton Keynes, MK11 3LW, UK
UKHW020139200926
475702UK00005B/817
9781138045859